DCC

FOR
RAILWAY
MODELLERS

An area of Peter Annison's layout Martin's Creek, a logging railway.

DCC
FOR
RAILWAY
MODELLERS

FIONA FORTY

THE CROWOOD PRESS

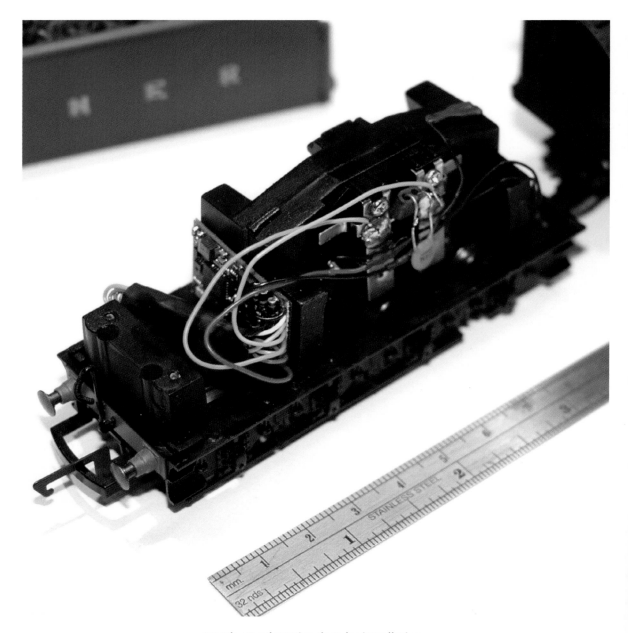

Hornby Tender Drive decoder installation.

CONTENTS

1 History of Digital Command Control and the NMRA   6

2 Digital Command Control   11

3 Common Misconceptions and Myths   17

4 Wiring the Layout   28

5 Track Design   33

6 The Basics of Wiring and Soldering   47

7 Locomotive Maintenance   62

8 The Basics of DCC Systems   80

9 Locomotive Decoders – DCC and DCC Sound   86

10 Fitting Decoders   105

11 Accessory Decoders   124

12 General Troubleshooting   128

Appendix   134

Glossary   152

Index   159

HISTORY OF DIGITAL COMMAND CONTROL AND THE NMRA

Digital Command Control (DCC) is a system for the digital operation of model railways. When equipped with Digital Command Control, locomotives on the same electrical section of track can be independently controlled. The DCC protocol is defined by the National Model Railroad Association (NMRA) and overseen by its Digital Command Control Working Group.

Early Systems

It may be surprising to learn that a version of DCC has been around since the 1940s, when Lionel Trains introduced a two-channel system called Magic Electrol. This used frequency control via a device in the locomotive called the E-Unit, and allowed two locomotives to run at the same time in opposite directions on the same track. In the 1960s, GE introduced a five-channel control system known as Automatic Simultaneous Train Controls (ASTRAC for short). This allowed five locomotives to run at the same time on the same track.

In the 1970s and 1980s, several more companies jumped on the control-system bandwagon:

- Dynatrol CTC-16 by Power Systems Inc. was a 16-channel system that appeared in 1979. Based

Hornby Zero system setup.

Lionel Trains electronic control.

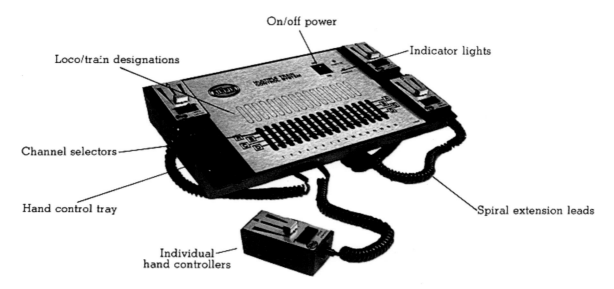

On/off power

Indicator lights

Loco/train designations

Channel selectors

Hand control tray

Spiral extension leads

Individual
hand controllers

Airfix multiple train control system.

on the Digitrack 1600 system, using integrated circuits, it could be built by following a series of articles published in *Model Railroader* magazine, beginning in December 1979.

- Hornby Zero-1 was a true digital system, introduced by the UK manufacturer Hornby to the USA in 1980 and Canada in 1981. Based around a Texas Instruments TMS1000, the system incorporated a four-bit microcontroller that was able to address and control up to sixteen locomotives.
- Airfix Multiple Train Control System (MTC) was an analogue-tuned carrier control system, introduced in 1979, which could control up to sixteen locomotives, with a maximum of four at a time.
- Zimo Digital began developing a digital command control system in 1977 and launched it in 1979. The major difference in comparison with Hornby's Zero-1 was that the Zimo system could control 99 trains and offered sixteen speed steps.
- Märklin Digital's system, which appeared in 1984, was designed for use with Märklin's line of Alternating Current three-rail HO locomotives. It was developed for Märklin by Lenz, using Motorola parts, hence the different mode and compatibility settings.

- The Trix Selectrix digital system appeared in 1982 and was designed for use with Trix's Direct Current two-rail locomotives.

Controlling Speed and Direction

Traditionally, the speed and direction of a model train have been controlled by varying the voltage and polarity on the rails. The higher the voltage, the faster the locomotive moves; the lower the voltage, the slower the locomotive moves. If the right rail is positive with respect to the left rail, the locomotive moves forward; if it is negative, the locomotive moves in reverse.

Being able to control the speed and direction of a train is great, but controlling more than one at a time is even better. Over the years, modellers have come up with many ingenious methods to achieve this. The basic one has been block wiring, in which the model railway layout is divided up into electrical blocks, each of which can control one locomotive. A cab (or throttle) is used to control

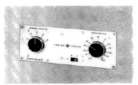

GE ASTRAC – a giant step forward in model railroading.

each train and arrays of selector switches connect each block. This method of control is also called cab control.

Probably the most ingenious method of cab control is also known as progressive cab control. As a train moves around the model railway, the connection between the cab and the block is automatically switched by relays to the next block, and the present block is released for another train to use. For a small layout, with one or two trains, block control is simple and straightforward to wire and instal. For bigger model railway layouts, however, the task can be immense.

The next evolutionary step is command control – a method of controlling individual locomotives (or trains) at the same time on the same rails.

The NMRA and the DCC Working Group

DCC has been around for many years in a range of formats, and today the main principles of the system are controlled by the NMRA so that everything is compatible. Before the 1930s, there were no common standards pertaining to any model railway equipment. Equipment supplied by one manufacturer would not necessarily work with that of another manufacturer, or even run on someone else's track. In addition, because many modellers built to their own standard or from their own designs and ideas, it was difficult, if not impossible, to take a locomotive to another modeller's railway and expect it to run without any issues. There were nearly as many couplers as there were manufacturers. It was a situation that was bound to be detrimental to the development of the hobby.

The National Model Railroad Association, or NMRA, came into being in 1935, when a gathering of model railroaders, manufacturers and publishers got together with the aim of bringing order to the chaos. The NMRA standards were developed to help ensure that equipment could be interchanged between one model railroad and another, and that carriages and locomotives of one manufacturer could run on the track of another manufacturer together with carriages and equipment of still other manufacturers and modellers.

Many of these basic standards have remained virtually unchanged from the time of their original publication in 1935. There have been some additions and refinements, but generally they have stood the test of time, proving to be of great benefit to model railroading. Their contribution to the development of the hobby to the point where it is today has been invaluable.

The first command control system was known as ASTRAC, developed by General Electric in 1964. As the electronics industry grew, new methods of controlling model trains were developed. Two of the most popular systems were Keller Engineering's Onboard, and PSI's Dynatrol, which both used audio tones to control each locomotive. Both systems worked well, but the user was still limited in the number of trains that could be controlled.

In 1978, *Model Railroader* magazine published a series of articles by Keith Gutierrez, the founder of CVP Products, giving readers instructions on how

to build their own command and control system. Called the CTC-16, it could control up to sixteen different trains, all on the same track. Many other companies went on to use the same methodology to control 32 or 64 trains.

The problem with all the systems that were being developed and built was the lack of standardisation. There was no common ground between them (except for the CTC-16), so, in the late 1980s, the NMRA set up a DCC Working Group to investigate the establishment of a standard. Rather than reinvent the wheel, the group decided to study all commercially available command and control systems, along with proposals received from Keller and Märklin. Their conclusion was that the best system on which to base the new standard was one that had been invented by Lenz Elektronik, which was used at that time by Märklin for their 2-rail sets. This system offered the best signalling method electrically, as well as the fewest limitations on expansion. The working group expanded the design, allowing for control of ten thousand possible locomotives, points and multiple-unit consists.

The group's revised standards were presented in 1993 and, by July 1994, they had been approved and adopted as the official NMRA DCC Standards. The finalised set of Recommended Practices was issued in 1995.

The DCC Working Group continues to clarify and expand the existing standards and recommended practices as the need arises – for example, with extensions such as Railcom.

2

DIGITAL COMMAND CONTROL

What Is Digital Command Control (DCC)?

In a nutshell, digital command control (also sometimes referred to as 'digital command and control') is a system, governed by a standard set by the NMRA, which

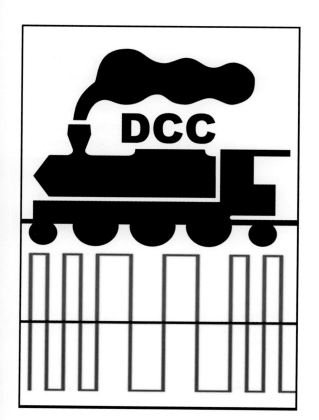

Digital command control (DCC).

offers the user the capability to run several trains at the same time, with the bonus of more simplified wiring.

As soon as the command station is switched on, the track receives full voltage. Changes to driving speed and direction are made by sending a signal to the decoders in the locomotives, which in turn control the locomotives. The track voltage is effectively AC, so direction is no longer dependent on polarity. Locomotives will respond only to those commands that are directly addressed to them. With DCC, the locomotives are being controlled by the user and not via control of the track voltage (as was the case with analogue control). When using a DCC system, full voltage is supplied all the time the command station is switched on. A controller is used to send information to the command station, telling it what operation a particular locomotive should do. The command station then transforms this information into a stream of digital code and sends it as packets via the track to the decoder, which instructs the locomotive to carry out the instruction as requested. Rather like a digital television, the locomotive will respond only to signals that are addressed directly to it and will ignore all other signals.

The same process applies for accessories, which can also be controlled by a decoder. Accessory decoders can be used to control points, lighting, turntables, servos for crossing gates, and much more.

There are many different terms used in the realm of DCC; a glossary is included at the end of the book to help with understanding.

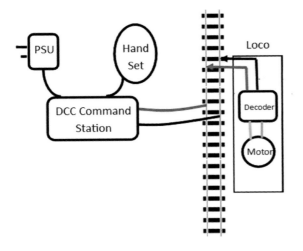

Basic DCC connections.

Choosing a System

When selecting a DCC system, asking a few questions will help you find your way around the range available:

- How many locomotives will be running at the same time?
- What gauge of locomotive will be run with the system?
- Will there be the control via DCC of points and/or signals to consider?

- Will computer control be required now or in the future?
- What power will be required for additional track-powered items, such as LEDs for carriage lighting, and so on.

The answers to these questions will point you towards identifying a system that will be the best fit. They can be used as a basis for discussions with a local retailer, who should be able to indicate what options are available, or for an online search. For more details on choosing a DCC system, *see* Chapter 8.

The Components of DCC

The basic requirements for a DCC system are a command station and handset, together with a suitable power supply. In addition, decoders will be needed for the locomotives. If points and signals are also to be controlled digitally, an accessory decoder will be required (*see* Chapter 11 for further details).

The choice of systems is wide-ranging, but, despite the many and varied opinions you will encounter among fellow modellers and journalists, there is no such thing as the 'best' one. The main advice should be to choose the system that you feel most comfortable with, and always try before you buy. All systems do the same thing, just in a slightly different manner.

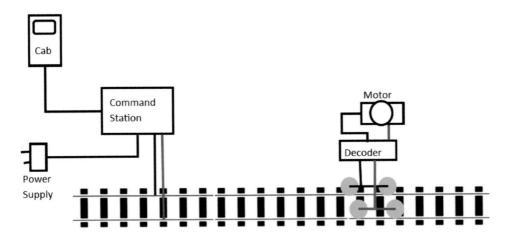

Basic components for DCC.

As a user, it is essential to be comfortable with how the handset feels in the hand, and with the activation of the buttons, knobs, and so on. Basically, if there is anything about the system that does not feel good, running the layout will not be an enjoyable experience.

A DCC system can be made up of either a desktop command station with a power supply or a command station with a handset and power supply. Desktop units may also have handsets added to them if required.

Command Station

Command stations come in all shapes and sizes, some requiring a handset, some not. Some allow for the use of a smartphone or tablet as the handset.

Handset

Handsets also come in all shapes and sizes and can be added as required, as command stations are able to operate with multiple handsets. The user's smartphone or tablet may also be used and, in some cases, can operate in conjunction with a DCC handset.

Power Supply

One aspect of a DCC system that is often forgotten is the power supply. It must suit the system purchased and, of course, it needs to be appropriate for the country in which it is to be used. These days, many manufacturers will provide a suitable power supply for their systems. However, if this is not the case, it is essential to have the right information -- the maximum voltage the system will take and whether it will take DC or AC – so that you can choose the correct power supply or transformer for the system.

Boosters

A booster for a DCC system is not a transformer or a power supply, but in principle an amplifier for the DCC signals from the command station. These are then combined with additional power output and the signals and power are sent to the track. Most DCC command stations come with boosters built in, and the total current output will be indicated in the manufacturer's details. Additional boosters may be required at a later date, depending on a change in the current requirements of the layout. Should the total power required by the layout exceed that of the

DCC command stations.

DCC handsets to go
with the command
stations.

Power supplies and
transformers.

chosen DCC system, perhaps because the layout has developed and grown, then an additional booster can be added to the layout.

Boosters are responsible for converting the AC or DC power from the power supply into local DC power to the track. They are also responsible for converting the signals from the command station into packets of information to be sent to the locomotives.

Different boosters may offer various features: short-circuit protection, automatic circuit breaker function, regulated voltage provision and auto reversing. These are not essential but can be added bonuses for the layout.

Boosters can be beneficial to a layout if it is somewhat more complicated or if it is being used for demonstration purposes, for example, at a club or exhibition. For the average home layout, however, a booster is not necessarily required, especially if best practices are followed for the wiring and the system has sufficient current for its running requirements.

Decoder

The next component required is a decoder for the locomotive, if it is not already fitted with one. Again, there are many makes, with different sizes and capabilities. To decide which is the best one

for your requirements, ask yourself the following questions:

- Is it just for control of the locomotive or is sound going to be wanted too?
- What is the current draw of the locomotive?
- Will there be a need to have operational lights on the locomotive? If yes, how many?
- Is the locomotive DCC-ready? If so, what socket does it have fitted?
- What space is there in the locomotive for a decoder? If sound is going to be fitted, then it is essential to consider the space required for the speaker too.

Some manufacturers may specify a particular decoder to be used in their locomotives. Although this is useful information for the user, it may be the case that a third-party decoder offers more features.

For more on decoder choices, *see* Chapter 9.

Accessory Decoder

The final optional component is an accessory decoder, which will allow for the digital operation of points, signals, lights, and much more.

Accessory decoders are normally multi-output units that are mounted under the layout in a stationary position. They can be used to control points/

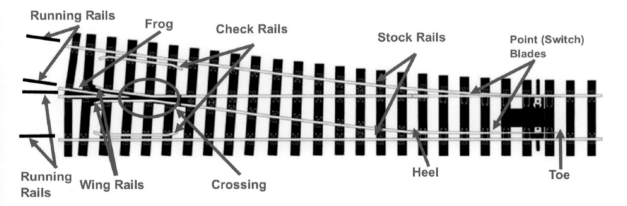

Anatomy of a Peco turnout/point.

turnouts, structure lights, scene lights, animation, signals, and other items that require an on/off signal. They may or may not provide power to the accessory.

As with everything else, accessory decoders come in a variety of shapes, sizes and outputs. They are usually multifunctional; however, when considering them for control of points, it is very important to know what point motors are being used. There are specific modules for solenoid (snap-action) point motors and for slow-action (stall motor) motors, as well as for servo motors.

Points (also known as turnouts and switches) are movable sections of track that allow a train to move from one line to another, guiding the wheels towards either the straight or the diverging track. In the model railway arena, points are used to change the direction of travel. They are typically made from metal and plastic and can be operated manually or by a point motor. The most important thing to remember when using points is that they must be correctly aligned with the track.

The term 'turnout' may be used to describe a point in some instances, but it is actually a combination of a number of components, only one of which is the actual point. The component known as the point is the short section of rail that physically moves to direct the train one way or the other. It is often referred to in model form as the 'point blades'.

Sig-naTrak ACE

Lenz Set 101

ECoS 50210

GM Prodigy Advanced

Digitrax DCS52

Digitrax EVOX

Hornby Elite

NCE Pro 5Amp

Roco Z21 with Multimaus

There can be many components to a DCC system, but only the command station, handset (if there is one), power supply and locomotive decoder are essential. All other items are optional.

3

COMMON MISCONCEPTIONS AND MYTHS

There are many misconceptions and myths around regarding DCC. Some of them are relevant and others are not. The main advice is always to ask as many questions as possible when starting out and do not believe everything you hear.

Misconceptions

'It's Expensive'

The thinking is that this type of control must be costly, even for the most basic system, but this is not necessarily the case. It is possible to start with a set-up that will do everything needed to control locomotives and even accessories for around £200. It is true, however, that a digital system that can control multiple locomotives at the same time, which will require either a multi-channel DC controller or individual controllers for each track, may cost less than a comparable analogue DC system. There are also continuous costs for locomotives that are to be run on a DCC layout, as they will all need decoders. A locomotive with no decoder fitted will not run on a DCC layout; it will just sit there and hum, with the motor oscillating very quickly. There are some systems that will allow the running of an analogue locomotive on Address 0, but these permit the user to control speed only. Again, the locomotive may produce a hum from the motor.

'It's Difficult'

There is a belief that you need to have a deep understanding of computers and computing. This is not the case at all. Setting up the command control system is fairly easy: two wires from the track to the command station, and two wires for power into the command, and everything is ready to go. The most difficult aspect is adjusting to a system that is different from the old DC one and learning how to control multiple locomotives at the same time. Although DCC systems are based on computer technology, there is no need to be an expert or even know anything about computers. However, although the actual use of the DCC system is very simple, installing decoders into the locomotive could present more of a challenge, especially in N gauge, and require some electrical knowledge. *See* Chapter 10 for further information.

'It's Intimidating'

Some people believe that DCC systems are too sophisticated and have too many buttons. They do come in all shapes and sizes, so a newcomer to the hobby may find them intimidating. Reading through the manuals – which have usually been written by technical engineers – can also be quite confusing. It is usually best to have a quick initial read-through, then go away and have a cup of tea before having a more detailed look. Most manufacturers also supply a 'quick-start guide' on running locomotives, and in principle it should be very simple to get going by following this. As progress is made and the need to delve deeper into the capabilities of the system arises, you can then sit down and read through the manual in detail or consider attending a course or workshop.

DCC is actually a true 'plug-and-play' product, which does what it says on the tin. Using it requires no special skills and it is ready to go straight out of the box.

It's a Niche Area of the Hobby

Command control has been around for over 40 years, and various versions of DCC have been available for over 20 years, but there are still some people who believe that it will never catch on! More and more layouts on the exhibition circuit are now being run using DCC, giving visitors the opportunity to see the systems in action. As DCC becomes increasingly familiar and more popular, hobbyists are seeing the benefits of using it on their own layouts and getting into the DCC scene.

As with everything, DCC is not for everyone. Everyone has their own likes and dislikes and ways of doing things. Deciding to use DCC is a personal choice and as such everyone can make up their own mind as to what is best for them, once all the facts are to hand. Again, it may be worthwhile considering attending a course/workshop to get a better understanding on what it is all about.

Myths: True or False?

There are many myths and beliefs surrounding the different elements of DCC control; some are partly true, some are partly false, and some are just plain wrong!

Signals

DCC track signal is AC or DC
Partly FALSE. Although it is like AC, it is actually neither AC nor DC. DCC is digital data sent in the form of pulse width modulation (PWM) on the rails. If the signal were to be looked at through an oscilloscope, a square wave would be seen.

DCC uses a carrier signal
FALSE. DCC does not use any carrier signals to transmit information. It is purely digital.

DCC uses DC with a signal riding on top to control the locomotive
FALSE. The DCC signal on the track is composed of both power and data. The signal is an AC-like digital waveform of a suitable amplitude to power the locomotives.

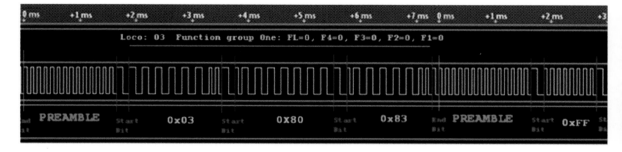

DCC square wave as seen through an oscilloscope.

The DCC signal is a low-frequency square wave with a high-frequency control signal

FALSE. The DCC waveform is both power and data, where the binary data is encoded in the form of long (100µSecond) or short (58µS) pulses. The DCC signal will vary between 5000 and 9000 Hertz (Hz) because of the pulse widths switching between those two states.

The DCC signal has polarity

FALSE. There is no concept of polarity with DCC. DCC is phased. A phase mismatch causes a short-circuit. The NMRA's DCC Standard states that the rail considered positive is impossible to define, so, when talking about wiring, the rails are referred to as right and left or red (+ve) and black (-ve).

DCC uses a differential signal on the rails

FALSE. While the NMRA DCC Standard describes the signal as a differential signal with no ground, this is a description only of what the waveform looks like on an oscilloscope, not of the signals themselves. Differential signalling requires two complementary signals, which are then summed to eliminate/reduce noise-induced errors when using high-speed communications between devices. As DCC uses a much higher voltage with a much slower data rate, it is much less susceptible to noise. The DCC protocol also includes error detection.

Voltages

There are positive and negative voltages on the track

FALSE. On a DCC track there are only positive voltages present. The phase of the rails does not control the direction as with analogue, hence there is no polarity. This then means that there are only two possible states: high or low.

DCC needs a negative voltage to reverse direction

FALSE. The multifunction decoder determines the direction, not the track voltage. Motor control is achieved by the switching sequence of the motor circuit.

DCC voltages can be correctly read with a multimeter

FALSE. Only a purpose-built DCC meter such as the RRamp meter or an oscilloscope will give accurate readings. This is also true for a true RMS meter. A regular digital meter set to AC volts will give only an approximate voltage.

Wiring

Analogue and DCC can be run on the same layout

FALSE. When direct current and DCC meet, only bad things can happen. If it is necessary to use analogue power, the layout can be wired so that only one source of power can be connected to the layout at any one time.

Only two wires are needed for DCC

Technically TRUE but in practice FALSE. Although DCC eliminates much of the wiring that was needed for analogue operations, and a very simple layout can be run quite happily with only two wires, it is not recommended.

Terminators must but used if the track bus is long

FALSE. Most DCC manufacturers do not insist on or recommend terminators. The track signals should be checked first to determine if there is a problem, which may require a different solution. Many signal integrity issues are directly related to inadequate wiring. Where there are very long lengths of wire, it may be advisable to twist the wires every so often to avoid interference.

Boosters

Boosters are brand-specific

FALSE. Many boosters can be interfaced to a different brand of command station. Most boosters have a low voltage or logic level input, while some can work with track voltage. The myth comes from the fact that it can be a bit of a challenge to determine precisely how to connect it. If, however, there is any doubt about connecting a booster correctly when adding it to a layout, it is advisable to use the same brand as the system.

DCC will not work without the addition of costly boosters

FALSE. Additional boosters provide additional power, as required, but are not always essential. Typically, a DCC system will include a built-in booster.

Larger layouts need additional boosters

FALSE. The number of boosters required is based on the power consumption, which depends on several factors. How many locomotives are running at the same time? Do they have sound or lights? Are lighted passenger carriages being run on the layout? Are there accessory decoders on the track bus? Has best practice been used in wiring the layout?

The boosters with higher current ratings are better

FALSE. An oversized booster will have too much available inrush current for smaller gauges, which could cause damage to a locomotive, unless circuit protection has been set correctly. If a power management device has been employed and the layout has been divided into power districts, with a lower current setting (for example, 4A), this may make sense. A high-current booster can deliver a significant amount of current into a circuit – as much as 60A for a brief period – which could result in damage before a circuit breaker has the chance to react to it.

The command station will shut down when a booster is shorted

MAYBE. If there is only one command station with a built-in booster and no circuit breaker at all, then this can happen. The protection in any booster or command station is designed to protect only the equipment of which it is a part. Other standalone boosters will continue as if nothing is wrong.

The exception to this rule is the NCE Power Cab. Unfortunately, due to its integrated design, a short will cause the entire unit to reboot. For this reason, NCE offers a protection module called the CP6, which has the role of limiting the current and preventing a reboot.

Decoders

Sound decoders need a lot more power

FALSE. High inrush current occurs only at cold start-up. Otherwise, sound only needs about 20% more power than a motor decoder. Adding an energy storage device to any decoder will also increase the inrush current.

BEMF (back electro motive force) never works in a consist

Not an absolute TRUE/FALSE situation. BEMF is sensitive to mismatches of decoder, locomotive or manufacturer and can be difficult to set up correctly. However, it can be done. Most users do not need BEMF, and it is acceptable to disable it if it is causing issues. Some multifunction decoders will disable BEMF when in a consist.

Programming on the main (OPS mode) is dangerous

FALSE. Programming on the main (POM) means the command station is instructed to send commands to change a specific CV to a specific decoder address. It is the same as sending a horn or light instruction. Blast mode, often initiated by using Address 0, causes instructions to be sent to every decoder in every locomotive on the rails. OPS and blast modes are not used with a dedicated programming track – this frequently leads to confusion. It is worth noting that there is no read-back of CVs in POM or OPS modes, unless Lenz's Railcom or Digitrax's Transponding technologies have been implemented into the layout and the decoders in the locomotives have Railcom/Transponding available and enabled.

All locomotives have to be converted to DCC

FALSE. Only those locomotives that will be used on a regular basis need to be converted to DCC. Some may not be easily converted, while others may not merit it. Also, if a non-DCC-ready locomotive has any particular value, this may be lost once it is converted to DCC.

Stall current is important

TRUE, but only for older and large-scale locomotives. Newer locomotives have improved drivetrains and motors that are more efficient and do not draw as much current. It is worthwhile checking. Most decoders will shut down if they overheat, and many modern decoders can handle any amp of current to the motor. It is however advisable to know the stall current for locomotives that are much older or non-DCC-ready, large-scale or kit-built, and those that have a motor for which the details are not known.

'DCC-ready' means the locomotive has a decoder

FALSE. The term 'DCC-ready' is used to identify a locomotive that can be converted easily to DCC. In most cases, it will have a socket ready wired in with a blanking plug in position to enable it to run on DC. It does not mean it has a decoder installed. 'DCC-fitted', on the other hand, does indicate that the locomotive already has a decoder.

The selection of a decoder is determined by the gauge

FALSE. Many N gauge decoders will work in an OO gauge locomotive without problems. It is possible to pay more and get fewer functions with an N or Z gauge decoder. The maximum current should always be checked against the motor and the size available for fitting the decoder.

DCC decoders will reset to their default address when running on a continuous loop

FALSE. This myth first appeared at train shows many years ago due to a lack of understanding at the time. Even though many exhibition and home layouts are loops, the decoder will never reset itself due to running in this format. It is possible for a decoder to be corrupted, but running in a loop will not be the cause.

A decoder testing device is essential

FALSE. While a decoder tester is a useful item, it is not a requirement. However, it may be a worthwhile investment when looking to do a decoder installation, for testing and programming beforehand.

CV29 is different among decoders

FALSE. The four mandatory NMRA CVs – CV1 Primary (Short) Address, CV7 Manufacturer Version Number, CV8 Manufacturer ID Number and CV29 Configuration Data – have to be the same for all decoders, regardless of manufacturer. CV29 has a different function and there may be slight differences between manufacturers' decoders, but the main features are the same.

General

Digital command control is beneficial only to a large layout

FALSE. DCC can benefit any layout, large or small.

DCC needs a computer to run

FALSE. There is no need whatsoever to have a computer to operate DCC. However, it could be a useful addition to a DCC system if automation is desired.

Programming requires a computer

FALSE. 'Programming' refers to the configuration of a decoder using the DCC system, either direct or via the handset.

The track must be perfectly clean or DCC will not work

FALSE. Although track must be kept reasonably clean for reliable operation, regardless of the control system used, the only time when it needs to be perfectly clean is when programming the locomotive on the programming track. If there is any dirt on the wheels or the track, it can interfere with the signals and the desired changes may not take place.

Points must be replaced with DCC-friendly ones

FALSE. There really is no such thing as a DCC-friendly point. If a point has worked fine with analogue operations, it will be effective with DCC as well. The issue is strictly electrical, and some points may require mechanical adjustments or modifications to eliminate any troublesome connection or improve operation.

Automotive tail-lights are necessary to protect against short-circuits

FALSE. In fact, they may defeat the short-circuit protection offered by the booster.

Wireless radio handsets for DCC are expensive

Not strictly TRUE or FALSE. A wireless radio-equipped handset will be considerably more expensive than a standard tethered unit, but there is a low-cost solution, using Wi-Fi devices with a JMRI (Java Model Railroad Interface) and a Wi-Fi hub. The computer along with an interface to the handset network can do the same thing. There are also standalone Wi-Fi devices available that connect to the handset network to offer the same capability.

Myths about Wiring

The myth that wiring for DCC has special technical requirements is perhaps the most common. If there is an existing DC controller connected to the layout or control panel in the conventional fashion with two wires (assuming there are no specialist circuitry or gadgets), then replacing it with a DCC set-up using the same two wires will work. It is as simple as that.

The key point is that there are no special skills required to get a DCC system up and running. It simply needs to be plugged in. Where trains used to run nicely, they will continue to run nicely, and problems will occur where problems used to occur. There will be no change. If there are any non-standard gadgets hooked up to the existing wiring, however – such as high-frequency track-cleaning units, like Relco, or high-frequency lighting units – they will need to be removed, as they are not compatible with DCC. That is about all that needs to be considered.

The second common theory about wiring is that a layout using DCC can be set up with only two wires. Strictly speaking, this is not a myth, as it is indeed feasible. However, it is not necessarily advisable.

Except on the very simplest layouts, most track that the UK modeller will be using, from off-the-shelf Peco through to lovingly hand-built formations, will need some insulating gaps to avoid short-circuits. This will be the case regardless of whether the layout is to be set up with a DCC system or a conventional DC system. Feeding power to the track beyond these insulating gaps will require another couple of wires. Points, signals and various gadgets such as reversing loops or train detectors will all need extra wiring, too. Clearly, although a 2-wire scenario is possible, it is not very likely. Those sidings that have no power when using Insulfrog points will also benefit from a track feed, and locomotives will then be able to be controlled with points set against them.

Even so, one of the big advantages of DCC is a huge reduction in the amount of wiring that is required. All those switches that people used to use to swap between controllers or to isolate sections where locomotives might be parked are completely redundant, as are the lengths of wire associated with them. Additionally, points, signals and reversing loops are connected to devices located adjacent to the items concerned, involving only short wiring

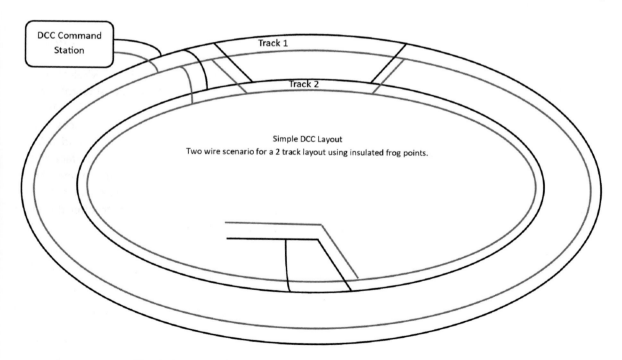

Simple layout with 2-wire connection from a DCC command system.

runs. Again, those miles of cable connecting back to control panels can become a thing of the past with DCC.

There is no need to wire a layout especially for DCC – if it has worked on DC, it will work on DCC – but there is one big difference: a DCC system can use a *lot* more power. As a guide, the average locomotive uses 0.3 amps, and the average DC controller supplies 1 amp. A DCC controller will supply 1 to 8 amps – which is enough to arc-weld with!

In short, although a DCC system does not require any special wiring techniques, it is vital to ensure that the wiring is up to the job.

Myths about Points

'Unfriendly' Points

Despite various opinions, there is no such thing as a DCC-unfriendly point (or a friendly one for that matter). Electrical rules are the same for both DC and DCC. If a layout works properly under one system, it will work properly under the other. It is that simple.

This myth probably derives from a few people experiencing issues with rogue models causing short-circuits. The improved short-circuit detection in DCC systems highlighted the problem, which users had not identified when running their layout in DC.

A short-circuit could occur this way in the event of a major, mechanical fault. Specifically, this would have to be a seriously out-of-gauge wheelset, which could then cause a short-circuit between two adjacent rails of different polarity – *if* (and only if) there are any on the layout.

Electrofrog Points

On Electrofrog points the two frog rails are metal all the way to the point. This means that the locomotive's wheels can pass over the frog without losing electrical contact. If the point is part of a loop or if there is a power feed on either of the tracks leading away from the frog, the frog will short the rails together. To avoid this, the power feeds from the frog need to be isolated with insulating joiners.

If one of the tracks is a siding with no power feed of its own, a normal joiner can be used, and the point will switch power to the siding, the same as an Insulfrog point.

Electrofrog points may be preferred because, being metal, they look more like the real thing. They also work much better, because there is no dead section of track for the trains to stall on. Live frog points are recommended for DCC, mainly because of there being no dead sections.

The disadvantages of using Electrofrog points are that they cost slightly more and that insulated joiners

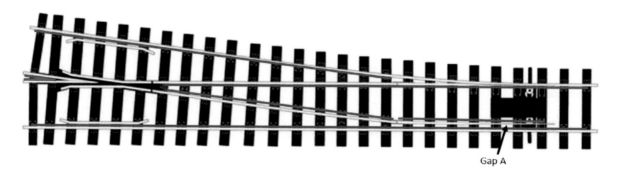

Gap A

An out-of-the-box Peco Electrofrog point, showing where an issue may occur. In an extreme case, an out-of-gauge wheelset could cause a short-circuit at gap 'A'.

are required. For Electrofrog single or double slips, switches will also be needed to change the polarity of the frogs.

For DCC it may be advisable (but not essential) to modify Peco Electrofrog points, and to use switches to change the frog polarity. This 'belt-and-braces approach' would only really be advisable if starting from scratch, when the track has yet to be laid down, or where there is a mixture of new and old rolling stock. In the latter case, the metal wheels may not be of the same gauge – for example, some may have thicker flanges than others.

An alternative to modification would be to add a point frog polarity-switching device, which operates by detecting the short-circuit created as a conductive wheel crosses on to the frog and then switches polarity.

Insulfrog vs Electrofrog

Insulfrog Turnouts/Points (aka Dead Frog):

- Diverging route is 'dead'.
- Insulation at nose of frog.
- Rail polarity is 'correct'.
- Power is switched by rails.

Electrofrog Turnouts/Points (aka Live Frog):

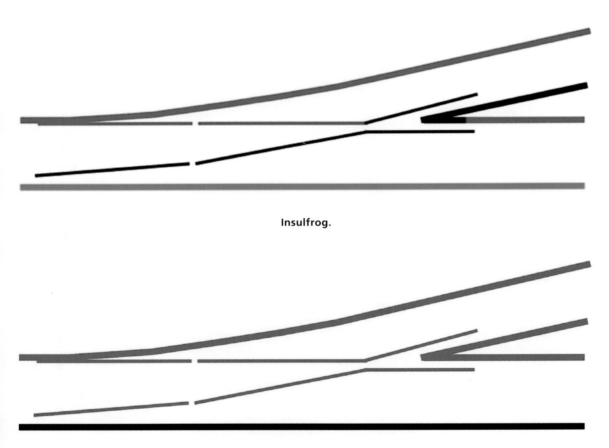

Insulfrog.

Electrofrog.

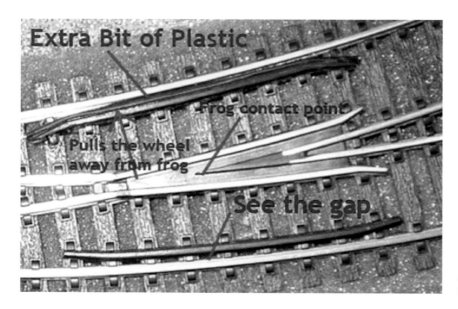

Adding plastic to the guide rail.

Understand your needs! There are various myths about DCC, but there is no need to worry too much, as long as you understand the requirements for your own layout and follow best practice.

- Diverging route is 'live'.
- Frog is uninsulated.
- Frog creates a short.
- Frog is switched by rails.

Fixing Points

To get rid of the unattractive hole in the layout where the point switching mechanism goes, cut a small piece of brown paper large enough to cover the holes in the baseboard. Make a slit in the middle to compensate for the operating rod, then sandwich the paper between the motor and the point. The operating rod needs to pass cleanly through the slit. The fixing lugs for the motor can be just pushed through the brown paper.

To stop the wheels riding up on to the frog and derailing, add a small piece of plastic to the guide rail. This will take the wheel away from the frog. The reason for adding the plastic to the inner guide rail is that the force of the locomotive cornering pushes the wheels towards the frog on the inside track and away from the frog on the outside track.

On Hornby points, the guide rails have no effect because the gap between the running rails and the guild rails is too big. Adding some plastic can make them start to work just like real points.

Make sure to wrap the plastic around the bends in the guide rails so that the wheels are persuaded sideways.

Express points are better and more realistic than standard points, but they are longer and almost twice the price. The 'S' shape that the standard points create when used with a second point to make a crossover is often too sharp and leads to derailment, especially in reverse and with long-wheelbase rolling stock. The express points have a much more gentle curve with a small straight in between. This straight section is important when setting out the track. 'S'-shaped sections should be eliminated to avoid buffer lock or uncoupling. If it is really necessary to include an 'S'-shaped section, it is very important to have at least a carriage's length in between the two bends. The longer the carriage, the longer the straight will have to be.

Using brown paper to cover the open hole.

4

WIRING THE LAYOUT

What to Consider

If there are a lot of circuits and accessories to instal under the baseboard, it may be better to mount them on vertical drop boards at the rear of the layout. It is a lot easier to work on boards like these than to work upside down on accessories mounted directly to the underside of the baseboard, especially if they can be unscrewed and taken somewhere more comfortable.

DCC has been designed to work even with atrocious wiring, but there is a big difference between adequate and optimal performance. To achieve the best performance, consider the following actions:

- Use separate power supplies or transformers for the command station, booster and accessories.
- If there is no power supply already included with the command station, select one with a voltage stability circuit. This will prevent the voltage dropping if there are lots of locomotives running at the same time.
- To estimate the total power requirement, calculate the maximum number of locomotives that

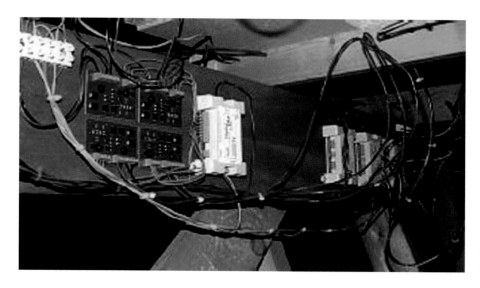

Drop-down board with accessory decoder.

WIRE SIZES AND CALCULATIONS

WIRE SIZES

In mains systems (240V AC), wire is sized primarily for amperage-carrying capacity (ampacity). The overriding concern is fire safety.

In low-voltage systems (12V DC), the overriding concern is power loss. Wire must not be sized merely for the ampacity, because there is less tolerance for voltage drop (except for very short runs). For example, at a constant wattage load, a 1V drop from 12V causes 20 times the power loss of a 1V drop from 240V.

Note: ampacity is based on the National Electrical Code (USA) for 30°C (85°F) ambient air temperature, for no more than three insulated conductors in an enclosed cable.

CALCULATION METHOD

This method works for any voltage or voltage drop, AWG (American Wire Gauge) or metric (mm²) sizing. It applies to typical DC circuits and to some simple AC circuits.

Step 1: Calculate the VDI (voltage drop index

VDI = (amps x feet) / (% volt drop x voltage)

Feet = the one-way wiring distance (1 metre = 3.28 feet)

% volt drop = the choice of the acceptable voltage drop (for example, use 3 for 3%)

Step 2: Determine the appropriate wire size from the chart

Compare the calculated VDI with the VDI in the chart to determine the closest wire size.

Amps must not exceed the ampacity indicated for the wire size.

Metric wire sizes are found using: VDI x 1.7 = mm²

Guide for commonly available mm sizes: 1/1.5/2.5/4/6/10/16/25/35/50/70/95/120 mm²

WORKING EXAMPLE

- 5-amp load 15V over a distance of 100 feet (30.48m) with 10% max. voltage drop
- VDI = (5 x 100) / (10 x 15) = 3.33
- For copper wire, the nearest VDI is 3
- This then indicates #12 AWG wire or 3.31mm²

Wire Size	Area	Copper wire	
AWG	MM²	VDI	Ampacity
16	1.31	1	10
14	2.08	2	15
12	3.31	3	20
10	5.26	5	30
8	8.37	8	55
6	13.3	12	75
4	21.1	20	95
2	33.6	31	130
0	53.5	49	170
00	67.4	62	195
000	85.0	78	225
0000	107	99	260

Wire size chart.

will run at any one time in an area and multiply this by *at least* 1.5. If the layout is large, it can be divided into logical operating areas (power districts), with one booster per district.

- Always ensure that the power supply has an amp rating that is larger than the controller can supply, to allow headroom for the power station to work properly.

Drop-down board removed and on the workbench.

Wiring Tips

Use the thickest bus wire around the layout that you can find and afford, and that will fit. This will ensure that the power can get to where it needs to be by the quickest possible route. Using the water analogy, the larger the tube, the more quickly the water will arrive at its end point. Using the largest possible wire means that the cable can carry more power (amps), which will in turn reduce voltage loss. Using household mains wiring can be very effective, as it can carry the necessary power with minimal voltage drop. Also, as mains wire has a single solid core, installation is easier.

Both wires should run parallel to each other. When working on long lengths (over 100 feet), an occasional twist in the wire will eliminate the possibility of AM radio interference close to the layout. It is possible that people may be advised to terminate or form the end of their power bus. However, in the model railway world, this is not essential, as we are not looking at the same number of data packets being sent along the bus and track between the DCC System and the DCC Decoder, which would require a termination, unless the total length of the track is over hundreds of metres. In the computing industry, terminations or forming is often used, as this will prevent signals from being lost or corrupted at the end of the transmission line due to the billions of data packets arriving at the same time.

There is also much talk about whether a bus should be continuous or end to end. Copper is conductive in any direction, so the shape of the bus in relation to the layout is irrelevant. It is advisable, however, to put it in a place where the dropper wires can be as short as possible, while still within reach of all the track areas required. Once the bus wires are in place, the installation of the dropper wires can be considered. These can be thinner than the main power bus wires if that makes them easier to instal, but, again, they should be as large as possible. It is also advisable to use flexible stranded wire. Droppers made from multi-stranded wire can be affixed to the main solid-core power bus by soldering, stripping off some of the insulation and then wrapping the stranded wire round, only using solder to fix it in position. Alternatively, suitcase connectors may be used.

There are no hard or fast rules for the placing of the droppers. Ideally, there would be a dropper for each individual piece of track, but in the real world this is not always possible. The advice is to set a regular distance, maybe 3 feet (1 metre), and put droppers at uniform intervals all around the layout, so

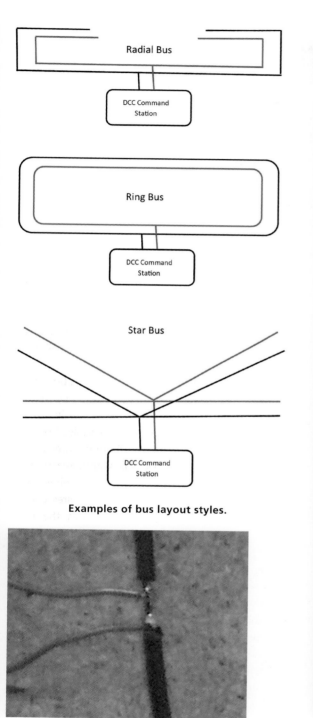

Radial Bus

DCC Command Station

Ring Bus

DCC Command Station

Star Bus

DCC Command Station

Examples of bus layout styles.

Example of dropper wire connected to bus.

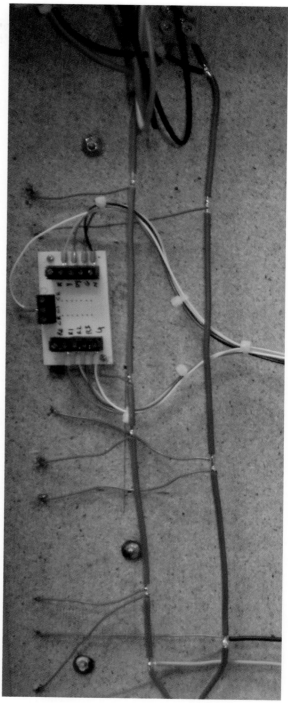

Example of bus with droppers in place.

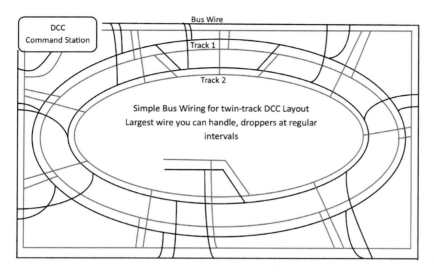

Example of bus layout complete with dropper wires.

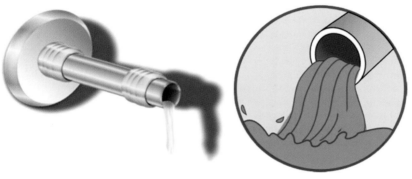

Small pipe, less water, slower flow

Large pipe, more water, faster flow

It helps to think of current as being like water. The bigger the water pipe, the quicker the water is delivered to where it is required, so the best principle for wiring a layout is to use the thickest wire that can be comfortably handled, for both the 'ring main' power bus and the droppers. The aim should be to distribute current evenly to all areas of the track and to have a uniform spread of wiring.

that there is a nice even spread of power. Of course, the distance between the droppers may vary due to turnouts or other track features being in the way.

If there are going to be many accessories along the track, such as lighting, point motors, signals, crossing gates, and so on, it is best to add a separate accessory power bus to the layout with its own power supply. This will allow the power from the DCC system to be saved solely for running and controlling the loco-motives and accessory decoders, if applicable.

The same principles apply for the bus wire and dropper wire as for the layout: the bus wire should be as thick as can be managed and the dropper wire a little thinner, but again as thick as is best for the modules and items to which it will be connected. It is essential to ensure that the power supply is suitable for all the items that are going to be connected to the bus. In other words, is there a need for a DC or an AC power supply, and how many amps are required? Always go for the most that can be derived from the power supply being considered. LEDs, for example, can draw on average 20mA each, so their demand can add up quickly and catch you out.

5

TRACK DESIGN

Gauge and Scale

In railway modelling, the term 'gauge' refers to the width between the rails being used by a scale. For example, OO gauge is the name given to modelling in 1:76 scale with a 16.5mm track width; N gauge refers to modelling in 1:160 scale with a 9mm gauge track width.

A drawing at a scale of 1:10 means that the object is reproduced ten times smaller than in real life (scale 1:1). As the numbers increase in the scale, the elements in the drawing get smaller. This is exactly the same for models.

Model railway scales and gauges for Europe were standardised by MOROP as NEM 010, which covers several gauges for each scale, published in both French and German with unofficial English translations. Narrow gauges are indicated by an

GAUGE AND SCALE

Gauge is the size of track in mm compared to the prototype.

Scale is the ratio size of an item in comparison to the full size item 1:1.

The most popular scales in the model railway world

Scale	Ratio	Standard gauge	M	E	I	P
T	1:145	3mm (0.118in)				
Z	1:220	6.5mm (0.256in)	4.5mm			
N (2mm)	1:160	97mm (0.354in)	6.5mm	4.5mm		
TT	1:120	12mm (0.472in)	9mm	6.5mm	4.5mm	
HO	1:87	16.5mm (0.65in)	12mm	9mm	6.5mm	4.5mm
OO (4mm)	1:76.2	16.5mm (0.65in)	12mm	9mm	6.5mm	4.5mm
EM	1:76	18.2mm (0.717in)				
P4	1:76.2	18.83mm (0.741in)				
S	1:64	22.5mm (0.886in)	16.5mm	12mm	9mm	6.5mm
O (7mm)	1:43.5	32mm (1.26in)	22.5mm	16.5mm	12mm	9mm
1	1:32	45mm (1.772in)	32mm	22.5mm	16.5mm	12mm
G	1:22.5	45mm (1.75in)				

additional letter added after the base scale, as follows:

- No letter = standard gauge (prototype: 1250–1700mm or 49.2–66.9in)
- M = metre gauge (prototype: 850–1250mm or 33.5–49.2 in)
- E = narrow gauge (prototype: 650–850mm or 25.6–33.5 in)
- I = industrial (prototype: 400–650mm or 15.7–25.6in)
- P = park railway (prototype: 300–400mm or 11.8–15.7in)

The NMRA standardised the first model railway scales in the 1940s and these were used widely in North America and worldwide. The NMRA and NEM Standards are compatible with each other. Full details can be found online.

Model Train Layouts: History and Inspiration

The hobby of railway modelling, as it is known in the UK, Australia, New Zealand and Ireland, or model railroading, as it is known in the USA and Canada, involves the representation of rail transport systems at a reduced scale. Scale models of a range of railway or railroad elements – locomotives, rolling stock, tracks, signalling equipment, cranes, and so on – are reproduced in rural or urban landscapes that can include features such as roads, bridges, buildings, vehicles, harbours, rivers, hills, tunnels and canyons. Many layouts are further enhanced by lighting and peopled by model figures. The possibilities for the enthusiastic modeller are endless!

The history of railway modelling dates back to the mid-nineteenth century, when miniature trains and track layouts were built as expensive toys for the children of wealthy families. One of the most famous of these was the Chemin de fer du Prince Impérial (the Railway of the Prince Imperial), created in 1859 for the 3-year-old son of Emperor Napoleon III. Sited in the grounds of the Château de Saint-Cloud in Paris, it was powered by clockwork and ran in a figure-of-eight. Half a century later, electric trains began to appear, but these were nowhere near as realistic as the model trains of today. There have been many technological advancements since the beginning of the twentieth century, and modern-day hobbyists are now able to create extensive layouts, controlling numerous locomotives and many other elements all at the same time.

Many railway modellers choose to recreate real locations and periods throughout history. There

Early metal model train.

Early train set.

**Early American
model train.**

are many sources of inspiration for this area of the hobby, including museums and other displays, as well as model railway shows that take place throughout the year. The world's oldest working model railway is housed today in the National Railway Museum in York, England. Designed to train signalmen on the Lancashire and Yorkshire Railway, it dates back to 1912 and remained in use until 1995. It was built by apprentices of the company's Horwich Works and supplied with rolling stock by Bassett-Lowke. Pendon Museum in Oxfordshire is also worth a visit, to see what is probably the largest model landscape in the UK – an EM gauge model of the Vale of White Horse in the 1930s. Pendon also houses one of the earliest scenic models, the Madder Valley layout built by John Ahern between the late 1930s and the late 1950s – a superb reflection of the developments that occurred in realistic modelling over that period.

A scenic view from the Vale Railway, housed in the Pendon Museum in Oxfordshire.

A small area of the Bekonscot model railway.

**Miniatur Wunderland –
Austria Viaduct.**

**Miniatur Wunderland –
Italy Vesuvius.**

Bassett-Lowke also supplied the parts and stock for the model railway of the miniature village at Bekonscot in Buckinghamshire, which dates from the 1930s. Covering 10 scale miles (450m) in length, it originally included two main-line circuits, with a branch line added later.

The world's largest model railway in H0 scale is the Miniatur Wunderland in Hamburg, Germany. First opened to the public in August 2001, it comprises eleven miniature 'worlds' and has a total track length of over 15,000m and more than 1000 trains. It was the brainchild of Frederik Brau, who was inspired by

Miniatur Wunderland – Sweden weather station.

Miniatur Wunderland – Switzerland Brichur Station.

a visit to a railway model shop in Zurich. The display that can be seen today is the result of hundreds of thousands of man hours of work and the investment of many millions of euros. It is extraordinarily detailed, with representations of entire countries, such as the USA, regions such as Patagonia and Central Germany, and cities, including Rio de Janeiro and Venice. Wunderland's fictional city of Knuffingen

even has a fully functioning airport with its own crowded check-in area and team of firefighters! It is certainly an inspiring place to visit.

The best advice when making decisions on a new layout, or on developing an existing one, is to visit one of the various museum displays or model railway shows and have a good look at the work of other modellers.

Choosing Locomotives

Before looking at what locomotives to buy, you will need to decide which is the most suitable gauge/ scale for the layout, in terms of the building process, the space available for the layout, and the ease of handling for the operator.

The next aspect to consider is which type of locomotive – steam, diesel, electric – will be run on the layout, and in which liveries. When it comes to making a purchase, go for the best quality you can afford, as these will stand the test of time. Check that all or most of the wheels have electrical pickup, especially for DCC running. If they are to be run using DCC, ensuring that the locomotives are DCC-ready will save a great deal of time when installing a decoder. If sound is going to be an option for the layout, it may be best to check out whether a sound-fitted locomotive is affordable – adding a sound decoder at a later date may cost as much as £100+.

When considering second-hand or graded loco-motives, always read the description in depth to understand exactly what is being offered and if in doubt, ask questions. Where possible, ask to see the locomotive running. Remember that, if an old loco-motive is modified to run with DCC, in some cases it will be difficult to return it back to DC analogue control.

One very important piece of advice is to buy what *you* think is best for *your* layout. Everyone has different tastes and aims when putting together a layout. Railway modelling is a hobby, which means that it exists for the pleasure of the operator, so the operator should buy whatever takes their fancy – if the budget allows. It is a mistake to think that you cannot buy an item because it does not 'go with' rest of the layout or the locomotives, or to be influenced by peer pressure. If you believe a locomotive or other piece of equipment will give you enjoyment in your hobby, it is definitely worth considering.

Track Design Considerations

Operation

DCC control has almost done away with the need to isolate sections electronically. As it gives you inde-pendent control of each locomotive, it allows you to have a siding full of locomotives and to make a single one move on its own at any time.

In the past, running two or more trains independ-ently of each other required a second controller and a second loop of track. If there was a need to cross from one loop to the other, the first train had to be moved into an area where it could be isolated (for example, a passing loop or a siding) before the second train could be moved on to the first train's loop. If required, the first train then needed to be moved on to the second train's loop. The whole process would then need to be repeated in order to have the two trains on the opposite tracks. Not only did the trains have to be moved into isolated sections, but it was also essential to remember to move the relevant switches in order

Use two coloured pens to draw the layout when starting out.

to move the trains. The isolated sections also needed to be big enough to accommodate all the trains.

DCC makes life much easier in this respect, making it unnecessary to think of the consequences of crossing the circuits by changing the points, or accidentally running two trains at the same time because the section has not been isolated or the wrong point has been switched. The only issue with DCC is the process of getting used to the fact that it is possible to run two trains head on into each other on the same piece of track. It is great for crashes!

When designing a track, it is essential to know whether there are likely to be any shorts created, and if so where. This is the case for both traditional and digital track operation. The best advice when starting out is to draw the layout using two coloured pens tied together – one colour (perhaps red) represents the right rail and another (perhaps black) the left. This will show exactly where shorts may occur, and in turn will help to work out the best option for eliminating the potential for any problems in these areas.

Reverse Loops

The wiring of reverse loops is the same for both analogue and digital layouts. The biggest difference is that no switches are required for a DCC-running set-up. It will need a reverse loop module, which will automatically change the polarity when crossing through the loop and thus keep the locomotive running.

Some DCC systems provide a detection circuit for automated reverse-loop operation. However, it is important to be aware that, when using such a system, any locomotive that is not equipped with a decoder may unexpectedly reverse its direction.

The use of automated reverse loops requires locomotives to have power pickup from both rails in multiple locations. Older locomotives may have pickups only from a single wheel and will need to have further pickups added. Also, by its very nature, a reverse loop reverses the power supply (see below right).

The diagram (see below left) shows an analogue layout with a reverse loop. The positive (red) ends up on the negative and the negative (green) ends up on the positive. This would result in a short-circuit. To get around this issue, two isolating sections need to be put in place. There is always a need to have one open and one closed depending on the direction of the points. If the points were set to straight, the locomotive could be driven in and around the loop until just before the top isolating section. (It is a good idea to put a signal here, operated by the same switch as the isolation section.) Once the locomotive has stopped, the bottom isolation switch is changed to closed, and then the top switch to open. The points must also be changed to the open position. Now the train can be driven out – the power supply has been switched, so that the direction in which the train

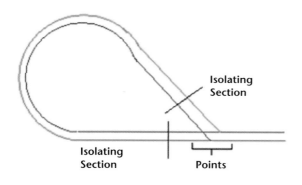

Typical reverse loop.

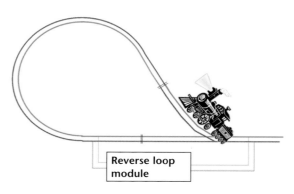

Reverse loop in DCC – the locomotive crosses the points and approaches the isolated section.

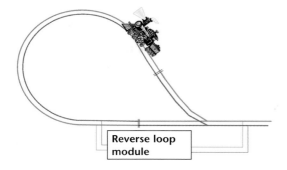

Reverse loop in DCC – the locomotive crosses into the isolated section.

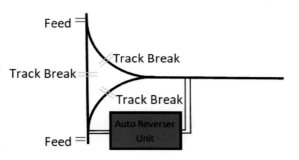

Diagram of a wye reverse loop.

was powered in will now power the train backwards and not forwards.

In a DCC system, the switches would be replaced with an automatic reverse loop module (or similar), which would automatically change over the power supply, thus avoiding short-circuits and keeping the train running through the loop without interference or any kind of manual switching.

Always remember that reversing sections can also be 'wyes', reversing triangles or turntables and not just the normal reverse loop. This is another good reason to draw the layout early on using the two-pen method. It is essential to ensure that the polarity of the track is consistent and, by using automated reverse loop modules, the track polarity can be switched under the train without any consequences, such as a change of direction.

With a wye, either leg of a wye or a tail can be used as the reversing section. These can be quite common on full-sized railways where space is not an issue. The triangle may connect to a separate line or may simply be a convenient spur siding.

Space and Shape

Another important aspect to consider is the amount of space that is available for the layout. What is the maximum curve that can be incorporated? The answer to this will dictate which locomotives will work best on the layout. Manufacturers will usually advise that the minimum radius for curves is radius 2. Anything smaller than this will lead to derailments. It is worth noting that many manufacturers do not design their models to run on first radius curves, including the deviating curve on small radius points.

There are many different types of modeller, from collectors of model trains, who wish to invest time and energy in building a landscape for their trains to pass through, to those whose aim is to create a layout based on a prototypical railway, to be operated as if it were a real railway system. Modellers who choose to model a prototype may want to build track-by-track reproductions of the real railway in miniature, often using prototype track diagrams and historic maps.

Layouts also vary widely in design, from a circle or oval of track to a realistic reproduction of a real

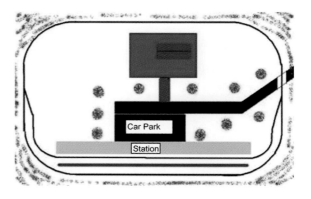

Example of possible high-speed layout design.

place, modelled to scale. The shape and design of the eventual layout will depend on the aims of the creator and on the type of trains that are involved. For long express trains, for example, it is best to have as much straight track as possible, as well as bends with a large radius. This will make the express trains look more appropriate when they are running at high speeds and also allow them to run safely at those speeds. More modern locomotives are much more powerful, so they can pull ever-longer loads faster. Radius 3 track is recommended for high-speed and long-wheelbase locomotives and carriages.

For small locomotives and wagons such as tank engines pulling coal trucks, a curvier layout might be required, with the track going in and out of valleys, through tunnels and over bridges. These trains should be run slower and due to the size of the carriages and loading gauges are smaller on bends than long carriages. A small slow engine weaving in and out of features looks more appropriate than a large express doing the same. The same goes for climbing up and down. Also because of the small size of the carriages, the designated minimum of a 'carriage length' in the middle of an S-bend is much smaller, or even non-existent. This means derailment is unlikely.

Points to Note

In summary, when designing a layout, it will help to note the following points before starting any building:

- The locomotive, not the track, is being controlled, so there are fewer switches. Two trains can run on the same piece of track in different directions and multiple locomotives can be running at the same time.
- DCC can be used to control lights independently in both locomotives and rolling stock.
- It is essential to ensure that there is sufficient power to the track and that it is uniformly spread to facilitate a good clean signal for the decoder.
- Accessory decoders can control fixed layout equipment such as points, signals, crossings, building lights, and so on.
- For easier access to the various modules, consider putting them on a drop-down panel, which could also be removed if required, to work on them on a bench.
- There is independent control of multiple locomotives on the layout at the same time. Providing there is sufficient amperage (current), there can be several locomotives running at the same time. The guideline for N and OO gauge is 0.3A per locomotive, so a 3-amp system should

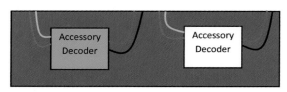

Red & Black Wires = Track connections

Yellow & Green Wires = accessory connections

Drop-down accessory decoder panel in situ.

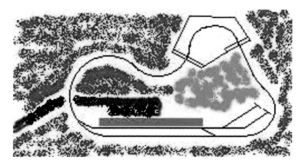

Example of possible small layout design.

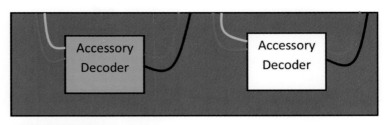

Red & Black Wires = Track connections

Yellow & Green Wires = accessory connections

Drop-down accessory decoder panel removed and put on the workbench.

Demonstration/testing layout at DCC Supplies. Covering three scales – N, OO and O – it is used to video locomotives once they have been repaired or had DCC or sound fitted, so it has been enhanced with scenery. (Published with kind permission of DCC Supplies; particular thanks to Richard, who spent his spare time putting the scenes together)

allow at least ten locomotives to be run at the same time.

- Lights in the locomotives and carriages will draw extra current from the track.
- Although it is possible to wire a layout with just two wires, it is not advisable. If a main power bus is used for the track, with droppers set at regular intervals, there should be a good spread of power around the layout.
- All dropper wires should be kept as short as is possible, so that the wiring under the layout is neat and clean.
- A separate power bus should be used for the trackside accessories, such as lighting, signals, accessory decoders, and so on.

There are a number of other factors to consider:

- Where is the layout going to be operated from?
- What system is being looked at and how many handsets are required?
- Is there a need to have the ability to walk freely around the layout?
- Is access required to all areas of the layout?
- Is the style of the layout to be aimed at a specific location, era or style?

Layout Sections

There is much discussion in the model railway world about the topic of layout sections and whether they are required or not. When talking about exhibition layouts, it is advisable to split the layout into sections as it is not ideal for the complete layout to shut down because of a short in one area. However, when talking about home layouts, then if best practice when wiring the layout has been followed, and all areas where there may be shorts created with points, crossings, etc. have been taken into consideration, the money required for sectioning the layout can be saved to be spent on locomotives, rolling stock or scenic items.

There are various reasons for splitting a layout into sections. First, it can assist with troubleshooting when shorts are detected. Only the section where the short is located will stop running, while the remainder of the track will keep going. If there is a lot of activity on the layout in various areas at the same time, which is using a lot of amperage, it may be more than the DCC system can handle. By splitting the layout into sections with boosters, the system can be kept running, with only one section being powered and DCC boosters being used to power the other sections.

Similarly, if the layout is quite complicated, and perhaps on different levels, splitting it into sections, each with their own boosters, will help greatly with power supply. It will also reduce short-circuit issues.

Scenery

For some people, the scenic side of the layout can be as important as the layout itself, and sometimes even more so. Certainly, once a scenic layout has been completed, it is very pleasing to see locomotives running through the landscape, whatever it may be.

The scenery of a layout can range from the bare minimum, with the base boards either painted or carpeted with static grass or scatter, a few ready-made buildings, and perhaps back scenes surrounding the outside walls, all the way through to a highly detailed landscape based on a real-life scenario. It can take years of work to get such a layout right.

When aiming to reproduce a real-life scenario, the best advice is to research it thoroughly, taking as many photographs as possible through every season of the year. If it is to be modelled on a certain period, and there is any historical interest in the scenario – perhaps cars, buildings or people – that will require more research. Once the information about the location, subject and time period has been gathered, the items for creating the diorama can be generated, either by purchasing off the shelf, or by building from scratch.

Various areas of Peter Annison's layout Martin's Creek, a logging railway.

There are many books and websites that can help with ideas for modelling the scenery that you would like to see on your layout.

Designed as a logging railway, Martin's Creek is a fine example of a home layout built to the design of the modeller Peter Annison. The scenery was built with whatever Peter had to hand to suit his needs. He describes how the layout came about:

The inspiration for my narrow-gauge logging model railway layout comes from years of bush walking searching for abandoned sawmills, tramways and bridges in the Australian bush. On my layout there are a few scenes of

Miniatur Wunderland (Italy). It should be possible to build a layout to suit you, your tastes, and the space available. Look at alternative options so that you can fit everything that you need into the space that you have.

places I have visited, and I have tried to replicate how I remember them. The millers used what materials were available nearby to build their structures; some were built from materials that were used in other locations until they relocated the mills to a new site near a forest. With that in mind, I have built

structures that look well used and in need of repairs. The whole layout is based on a fictitious tramway/railway almost broke and about to close due to the forests being logged, timber being scarce and there being a lack of funds for maintenance. It makes an interesting theme for a model railway.

One scenic area of Martin's Creek, Peter Annison's layout designed as a logging railway. (Published with kind permission of Peter Annison)

6

THE BASICS OF WIRING AND SOLDERING

Principles of Electricity

In order to appreciate the importance of doing a good job when soldering, it helps to understand the physics behind the practice.

Everything is made of atoms. An atom is a single part of an element, consisting of electrons, protons and neutrons. Electrons (-ve charge) are attracted to protons (+ve charge), and it is this charge that holds the atoms together. Some materials – air, glass, rubber, most plastics, and so on – have a strong attraction and refuse the loss of electrons; these are called insulators. On the other hand, some materials have a weak attraction, which allows electrons to be lost;

these are called conductors and include materials such as copper, silver, gold, aluminium, and more.

Electrons themselves can be made to move from one atom to another, creating an electrical current. A surplus of electrons is called a negative charge and

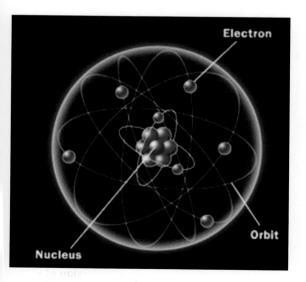

The atom is a single part of an element.

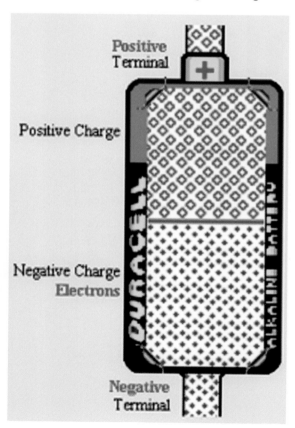

Cross-section of a battery.

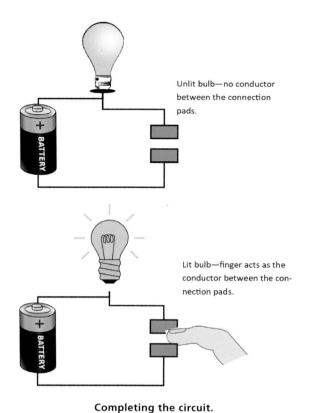

Unlit bulb—no conductor between the connection pads.

Lit bulb—finger acts as the conductor between the connection pads.

Completing the circuit.

a shortage of electrons is called a positive charge. A battery provides a surplus of electrons by means of a chemical reaction.

Connecting a conductor from the positive terminal to the negative will cause electrons to flow. This is illustrated in the school physics classroom where students learn that they can make a connection and light a bulb with one or two fingers. In this instance, the body acts as a conductor to complete the electrical circuit.

Current, Voltage and Resistance

The level of the electrical energy in a circuit can be measured in terms of both current and voltage. Current is the rate at which charge flows through a circuit; voltage is the strength of that charge at a given point.

A battery has a positive terminal and a negative terminal. The difference in charge between each terminal is the potential energy that the battery can provide. This energy is measured in units called volts.

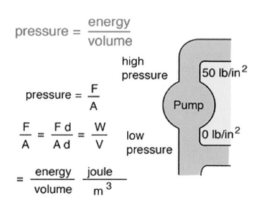

$$\text{pressure} = \frac{\text{energy}}{\text{volume}}$$

high pressure

50 lb/in^2

Pump

low pressure

0 lb/in^2

$$\text{pressure} = \frac{F}{A}$$

$$\frac{F}{A} = \frac{F\,d}{A\,d} = \frac{W}{V}$$

$$= \frac{\text{energy}}{\text{volume}} \quad \frac{\text{joule}}{m^3}$$

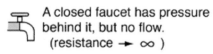

A closed faucet has pressure behind it, but no flow.
(resistance → ∞)

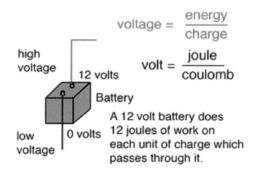

$$\text{voltage} = \frac{\text{energy}}{\text{charge}}$$

high voltage

12 volts

$$\text{volt} = \frac{\text{joule}}{\text{coulomb}}$$

Battery

low voltage

0 volts

A 12 volt battery does 12 joules of work on each unit of charge which passes through it.

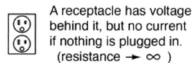

A receptacle has voltage behind it, but no current if nothing is plugged in.
(resistance → ∞)

Current represented by water.

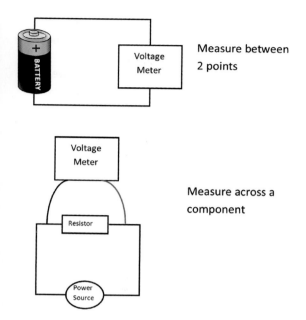

Measure between 2 points

Measure across a component

Measuring voltage.

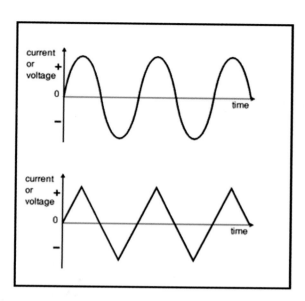

AC current.

Conventional flow notation

Electric charge moves from the positive (surplus) side of the battery to the negative (deficiency) side.

Conventional flow.

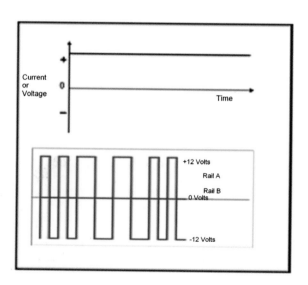

DC current.

A good analogy to help with understanding current and voltage is water running through a pipe. Current is represented by the volume of water flowing through the pipe per second; voltage is represented by the force of the water as it pushes past a given point.

Voltage is like differential pressure, so it is necessary to measure between two points or across a component in a circuit. When measuring DC voltage, it is essential to ensure that the polarity of the meter is correct: positive (+) is red, negative (-) is black.

Current on the other hand is the uniform flow of electrons through a circuit. Current is measured as the amount of charge (number of charged particles) flowing past a point in a circuit per second. It is measured in units of amperes (amps, or A), using an ammeter. There are two types of current: alternating current (AC) and direct current (DC). With alternating current, the voltage is continually changing between positive and negative. The rate of the changing direction is called the frequency, measured in hertz (Hz).

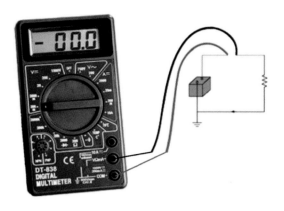

Measuring amps.

Constriction creates
Resistance to water flow

Resistor creates
Resistance to current flow

Resistance.

This is the number of forwards-backwards cycles per second. Mains electricity in the UK has a frequency of 50Hz.

An AC power supply is suitable for powering some devices such as lamps and heaters, but almost all electronic circuits require a steady DC power supply. With this in mind, it is essential to know which type

www.resistorguide.com

	Color	Signficant figures			Multiply	Tolerance (%)	Temp. Coeff. (ppm/K)	Fail Rate (%)
Bad	black	0	0	0	x 1		250 (U)	
Beer	brown	1	1	1	x 10	1 (F)	100 (S)	1
Rots	red	2	2	2	x 100	2 (G)	50 (R)	0.1
Our	orange	3	3	3	x 1K		15 (P)	0.01
Young	yellow	4	4	4	x 10K		25 (Q)	0.001
Guts	green	5	5	5	x 100K	0.5 (D)	20 (Z)	
But	blue	6	6	6	x 1M	0.25 (C)	10 (Z)	
Vodka	violet	7	7	7	x 10M	0.1 (B)	5 (M)	
Goes	grey	8	8	8	x 100M	0.05 (A)	1(K)	
Well	white	9	9	9	x 1G			
Get	gold			3rd digit only for 5 and 6 bands	x 0.1	5 (J)		
Some	silver				x 0.01	10 (K)		
Now!	none					20 (M)		

6 band — 3.21kΩ 1% 50ppm/K

5 band — 521Ω 1%

4 band — 82kΩ 5%

3 band — 330Ω 20%

gap between band 3 and 4 indicates reading direction

Resistor colour code chart.

of power supply all the modules and components on the layout require to power them.

Direct current always flows in the same direction, but it may increase and decrease. A DC voltage is always either positive or negative.

In order to measure the current, the circuit must be broken, and a meter installed between the device and the component. The measurement will be imperfect, however, because of the drop in voltage that will be created by the meter.

A level of resistance is provided by all materials. Measured in ohms, resistance depends on the cross-sectional area, material type and temperature. Power can be dissipated in the form of heat by a resistor, which is a non-directional component. There are many different types of resistor, with colour coding allowing identification without a meter.

Using a Multimeter

The multimeter is a very useful tool to have in the toolbox. It is a measurement tool for electronics that combines essential features such as a voltmeter, ohmmeter and ammeter and, in many cases, continuity. A multimeter can give an immediate understanding of what is going on with a circuit. Whenever something is not working,

A multimeter is a useful tool.

this tool will help with identifying the issue and troubleshooting:

- Is the switch on?
- Is the wire conducting electricity or is it broken?
- Is the solder joint conducting or is it a bad joint?
- How much current is flowing through this LED?
- What is the ohm reading of this resistor?
- How much power is left in the battery?

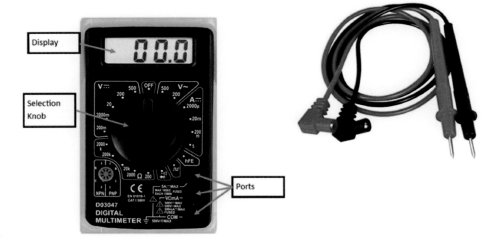

Multimeter sections.

There is a variety of multimeters on the market, ranging in cost from low to really quite high. There is no need to spend a fortune, as long as the multimeter is suitable for the task at hand. It is advisable to choose one that covers voltage, current, resistance and continuity. A continuity buzzer is particularly useful, as this will be a great help when trying to identify whether the connection is good, or non-existent, without needing to have direct sight of the meter.

There are three essential parts to the multimeter: the display, the selection knob and the ports, where the probes are plugged in (probes are typically red and black). Depending on the model of meter, there may be three or four ports. The convention for the colour coding of the probes is that black is usually the one connected to the COM port and red is then plugged into one of the other ports, depending on what is to be measured. For a typical multimeter with three ports, the set-up is as follows:

- COM: the black probe is plugged into this socket.
- 5A Max: this is used to measure large currents (greater than 200mA)
- VΩ: this is used to measure voltage and resistance and to test for continuity.

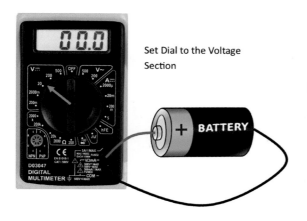

Set Dial to the Voltage Section

Measuring voltage.

Some meters, known as 'auto-ranging' types, adjust their range automatically, eliminating the need to change it manually.

Measuring Voltage

In terms of voltage, both DC and AC can be measured. The icon for DC is = and for AC ~. DCC voltages require an averaging RMS meter for a correct reading. It is still possible to read the voltage with a digital multimeter, and it will give a guide, but it will not be accurate.

To measure voltage of a component, the mode is set to the V section required (AC or DC). The red probe is plugged into the port with a V next to it, on the positive side of the component, as this is where the current is coming from. The black COM probe then goes to the other side of the component, and the value is read. Note: it is important to select the range that is nearer the expected voltage of the component being measured. For example, if the voltage of a 1.5V AA battery needs to be measured, the range is set at 2000mV. If there is any uncertainty as to what the voltage should be, it is advisable to go through the ranges. If 1 appears on the screen, then the voltage is higher than the range selected. If a higher range is selected, the value may be read, but with less accuracy.

A typical multimeter will have the following DC ranges:

- 200mV = 0.2V
- 2000mV = 2V
- 20V
- 200V
- 500V

Due to the risk of electrical shock, voltage measurement should be restricted to track voltages or below.

Voltage can only be tested when the circuit is powered or there will be nothing to test! It must be plugged in even if it does not seem to be working. Voltage is always measured between two points and both probes must be in circuit. When trying to test a point or read the voltage at this or that location, the black probe should be put at ground and the red (positive) probe at the point that is to be measured.

Voltage is directional. If a negative voltage is read in the circuit and it is certain that this cannot be the case, it will be necessary to check that the probes are the right way round.

Testing Resistance

Resistance can only be tested if the device being tested is not powered. Resistance testing works by powering the circuit and seeing how much current flows. If there is already voltage in the circuit, the readings will be incorrect.

Resistors can only be tested out of circuit. When they are in circuit, everything connected to the resistor will also be measured. Resistance is non-directional; the probes can be switched around and the reading will be the same.

Checking for Continuity

A circuit can be tested for continuity using a multi-meter with the appropriate setting. This will facilitate

Open circuit – no connection.

Continuous circuit with a connection.

easier identification of issues in the circuit, such as faulty wires or bad solder joints. This is one of the most important tests – not only does it check to see if a solder joint is good or a wire is broken, it can also be used to check whether there is no connection.

Checking for continuity.

Make sure that the multimeter is working for continuity by touching the two probe tips together. There should be a continuous sound, or the meter should read zero. With no power running through the circuit, use the multimeter to check across two points. If there is near-zero resistance between two points, they are electrically connected, and there will be a continuous sound from the multimeter. If there is a high resistance between two points (greater than approximately 10 ohms), there may be a wiring fault, unless another component is in circuit.

There is a fault with the connection if the meter indicates an open circuit (check the manual), there is no sound, or the sound is not continuous.

To check the continuity of a wire, connect each probe to the stripped wire ends. The same indicators will then apply.

Soldering

Soldering involves the use of a metal alloy to bond other metals together. Tin and lead are used in 'soft' solders, which melt rather easily at a lower temperature. The more tin there is in the alloy, the harder the solder and the higher the temperature required. Copper and zinc are used in 'hard' soldering, or brazing, which requires considerably more heat to melt.

Soldering electrical components requires different fluxes, solder and techniques from soldering metal kits. Soldering is not welding; the materials being joined do not melt and fuse together, and it cannot be used as a substitute for a good mechanical connection. The bond will break if it is stressed. A solder joint will not make a good electrical connection to dirty or corroded components or PCBs – the solder must 'wet' on to the surfaces being joined.

Although the concept of soldering is simple – join electrical parts together to form an electrical connection, using a molten mixture of solder (usually a mix of lead and tin) with a soldering iron – it is a delicate manual skill that comes only with practice.

The ability to solder effectively will determine directly how well the joints will last during their lifetime. Poor soldering can cause a major disappointment, which can damage confidence. The best advice for the beginner is to test skills on spare pieces of track, wire, stripboard or similar, to get used to the technique. Practice makes perfect!

Components of Soldering

Flux

Flux cleans the surfaces and promotes 'wetting'. There are various different types of flux on the market:

- Rosin flux: the most commonly used type, but it does leave a sticky residue that can only be cleaned off with a solvent (for example, isopropanol alcohol).
- Water-soluble flux: the residue can be removed with water.
- No-clean flux: the residue can be left on the board as it is inert and 'visually acceptable'.
- Acid fluxes: for use by plumbers only. They are not to be used in railway modelling.

Flux—Pens, Bottles, Syringes, Tubes or Solder filled with Flux

Fluxes.

Soldering Iron

There are two main types of soldering iron: a solder station, which is very useful for bench work, and a pencil-style iron, which is more useful for under-board work. A solder station will usually have temperature control and a base containing a transformer. Pencil-style irons do not normally have temperature control, but they are self-contained. Always make sure that the soldering iron selected will reach the required temperature for the solder that you plan to use.

Usually, a soldering iron will have interchangeable tips. This can be exceedingly useful as it allows you to choose the most suitable tip for each soldering job. When buying a soldering iron, ensure that tips are readily available. Conical tips have a fine, pointed end, and are ideal for smaller areas. A chisel tip is ideal for larger components due to the broad flat tip.

Along with the soldering iron, you will need a sponge or similar for cleaning the tip. One option is a wet sponge, which must be kept moist (not wet) at all times. Water can be kept to hand, preferably in a flexible bottle with a spout, to avoid spillages. The downside of using a damp sponge is that its use will reduce the lifespan of the tip, as the metal expands and contracts each time it is cleaned, due to the drop in temperature. The other option is a brass wire sponge, which cleans by abrasion. It does not require water, so it does not lower the temperature of the tip on application. However, tin-plated tips will need re-tinning after cleaning with this type of sponge.

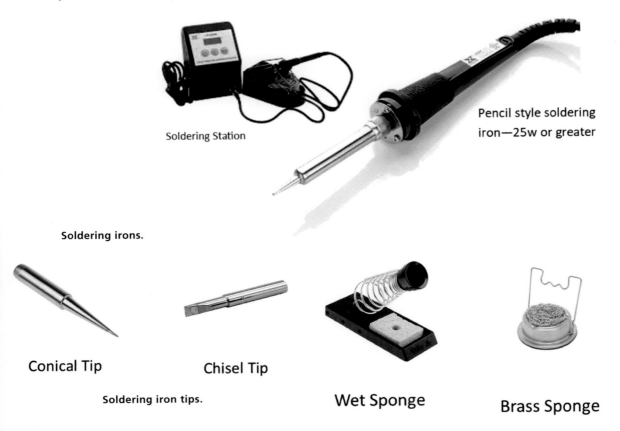

Soldering Station

Pencil style soldering iron—25w or greater

Soldering irons.

Conical Tip

Chisel Tip

Soldering iron tips.

Wet Sponge

Brass Sponge

Sponges for cleaning the tip of the soldering iron.

Soldering Iron Stand

This is a very basic tool, but very handy to have, as it prevents the hot iron tip coming into contact with flammable materials or causing accidental injury to skin. Most soldering stations come with a stand built in or included. The stand needs to be heavy enough not to fall over when the soldering iron is placed in it. A stand that has a space for the sponge is an added benefit.

Sturdy soldering iron stand with brass sponge.

Soldering iron stand with wet sponge.

Solder

Solder is the metal alloy material that is melted using the hot tip of the soldering iron to create the bond between electrical parts. It typically comes in both leaded and lead-free variations. Inside the core of the solder is flux, which helps improve the electrical contact and its mechanical strength.

Soldering iron stand with brass sponge.

Solder.

The material that is most commonly used in the electronics industry is lead-free rosin-core solder, usually made of a tin/copper alloy. There is also 60/40 leaded (60% tin, 40% lead) rosin-core solder available. Despite the rumours, leaded solder is still available for hobby use. However, it is banned for use in mass production in Europe. Lead-free solder has a higher melting point.

Help Hand (Third Hand)

A very useful device to give a helping hand! There are different variations out on the market. The most popular usually has 2 or more crocodile clips and a magnifying glass.

Desolder Pump (Solder Sucker)

Used for removing molten solder when a solder joint needs to be undone.

To use, press the plunger down at the end of the solder sucker, heat the joint then place the tip of the solder sucker over the hot solder and press the release button.

Desoldering Braid

Useful for removing molten solder when the joint needs to be undone. To use, place braid over the joint to be removed, heat the joint through the braid and pull the braid away quickly once the solder 'wets'.

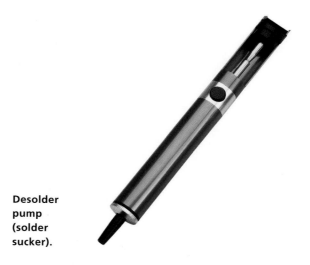

Desolder pump (solder sucker).

Desoldering braid.

Starting Soldering

The first step in soldering is to secure the work so that it does not move during the process, which could affect the accuracy. This can be achieved by use of a jig, a modeller's small vice, a multi-hand stand (also known as a 'third hand') or even a lump of blu-tack.

In some cases, solder joints may be required to have some degree of mechanical strength. For example, when a wire is soldered to a switch tag, the wire is looped through the tag and secured before solder is applied. The downside of this is that it will be more difficult to desolder if required. In the case of a circuit board or strip board, the wires of the components are bent to fit through the board and the legs splayed outwards a little before applying

Helping hand.

A good solder joint.

solder. This will help the part to grip the board. Excess lengths of leg or wire can be snipped away, but this should be done only after soldering as the extra length can be used to dissipate some of the heat away from the component.

The Principles of Soldering

A perfectly soldered joint will be nice and shiny-looking and will prove reliable in service. A little effort spent beforehand perfecting your soldering technique will save a considerable amount of time later, and avoid you having to troubleshoot.

There are a number of factors that are key to the successful soldering of a joint:

- cleanliness;
- temperature;
- time; and
- adequate solder coverage.

Cleanliness

The importance of cleanliness when soldering cannot be emphasised enough. First, all parts, without exception – including the iron tip itself – must be clean and free from contamination. Solder will not 'take' to a dirty part. Old components or copper board can be notoriously difficult to solder, because of the layer of oxidation that builds up on the surface of the leads and repels the molten solder. This will soon be evident because the solder will 'bead' into globules, going everywhere except where it is needed. Dirt is the enemy of a good-quality soldered joint!

It is an absolute necessity to ensure that all parts are free from grease, oxidation, and other contamination. In the case of old resistors or capacitors, for example, where the leads have started to oxidise, use a small hand-held file or knife blade, or rub a fine emery cloth over them to reveal fresh metal underneath. Stripboard and copper-printed circuit board will generally oxidise after a few months, especially if it has been fingerprinted. The copper strips can be cleaned using an abrasive rubber block, like an aggressive eraser, to reveal fresh shiny metal underneath. A glass-fibre brush is a useful tool here, used like a propelling pencil to remove any surface contamination.

After this type of freshening up, surfaces should be cleaned again using a cloth with some isopropanol

It cannot be said enough, the tip must be kept clean at all times during the soldering process. The only time the tip does not need to be cleaned is when the soldering iron is to be put away.

KEEP CALM AND KEEP THE TIP CLEAN

alcohol or similar, to remove grease marks and fingerprints. After this stage of preparation, it is important to avoid touching the parts again.

One side effect of having dirty surfaces is a tendency to apply more heat, to force the solder to take. This will often do more harm than good because it may not be possible to burn off any contaminants, and the component may overheat.

Before using the iron to make a joint, it should be prepared by 'tinning' (coating with solder) by applying a few millimetres of solder, then wiping on a damp sponge. This needs to be done straight away with a new bit, anyway. Reapplying a very small amount of solder again each time will improve the thermal contact between the iron and the joint, so that the solder will flow more quickly and easily. It is sometimes better also to tin wires or components to be soldered.

Temperature

Another tip for successful soldering is to ensure that the temperature of all the parts is raised to roughly the same level before applying solder. When trying to solder a resistor into place on a printed circuit board, for example, it is far better to heat both the copper PCB and the resistor lead at the same time before applying solder. This will allow the solder to flow much more readily over the joint. Heating one part but not the other will lead to a far less satisfactory joint. Make sure that the iron is in contact with all the components first, before touching the solder to it.

The melting point of most solder is in the region of 188°C (370°F) and the iron tip temperature is typically 330–350°C (626–662°F). The latest lead-free solders typically require a higher temperature.

Time

The joint should be heated with the tip for just the right amount of time, while a short length of solder is applied to the joint. Do not use the iron to carry molten solder over to the joint! The soldering iron is not a paintbrush, so do not use it as such.

Leaving it too long will damage the component and perhaps the circuit board copper foil too. Heat the joint with the tip of the iron, continue heating whilst applying solder, then remove the iron and allow the joint to cool. This should take only a few seconds for the more experienced individual. The

FIRST AID

In the event of a burn that requires treatment, follow these steps:

- Immediately cool the affected area with cold running water, ice, or even frozen peas for at least ten minutes.
- Remove jewellery before any swelling starts.
- Apply a sterile dressing to protect against infection.
- Do not apply lotions or creams, and do not prick any blisters that form later.
- Seek professional medical advice where necessary.

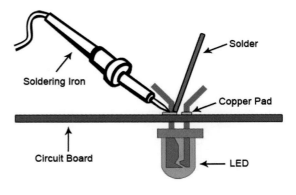

How to solder.

ACHIEVING THE PERFECT SOLDER JOINT

SUMMARY

- All parts must be clean and free from dirt and grease.
- Secure the work firmly, perhaps using a helping hand device.
- 'Tin' the iron tip by applying a small amount of solder. (Do this immediately with new tips when they are being used for the first time.)
- Clean the tip of the hot soldering iron on a sponge or similar.
- It helps to add a small amount of fresh solder to the cleansed tip.
- Heat all parts of the joint with the iron for under a second or so.
- Continue heating, then apply just enough solder to form an adequate joint.
- Remove and return the iron safely to its stand.
- It should take only 2 to 3 seconds at most to solder the average joint.
- Do not move parts until the solder has cooled.

TIPS AND TECHNIQUES

- Heat the wire with the iron.
- Touch the solder to the wire.
- Let the wire melt the solder (where possible).

- Keep the tip clean.
- Use the correct tip for the application.

SOLDERING THROUGH-HOLE COMPONENTS

- Ensure the component lead and the PCB land are clean, and the component is securely positioned.
- Clean the tip.
- Gently put the tip in contact with both the lead and the land.
- Touch the solder to the lead and land, not the tip. Let the solder flow into the joint and through the barrel.
- It may be an idea to add a little solder to the tip to get good thermal transfer to the lead and land.

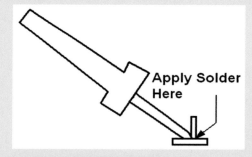

Apply solder here.

---

heating period depends on the temperature of the iron and the size of the joint. Larger parts also need more heat than smaller ones, but some parts are more sensitive to heat and should not be heated for more than a few seconds.

Solder Coverage

The final key factor in producing a successful solder joint is the appropriate amount of solder. Too much solder is an unnecessary waste and may cause short-circuits with adjacent joints. If there is too little solder, it may not support the component properly or may not fully form a working joint. Knowing how much to apply only really comes with practice.

Troubleshooting Guide

If the solder fails to 'take', there is probably some grease or dirt present. The solution is to desolder, clean up the parts properly and start again. The other possibility is that the material is not suitable for soldering with lead/tin solder (as would be the case with aluminium and zinc), or that the surface area of the intended joint is too large.

If the joint is crystalline or grainy-looking, it may be because the iron was moved before the joint had cooled sufficiently. It may also have happened because the joint was not heated adequately, perhaps due to the iron being too small or the joint being too large.

If the solder joint forms a 'spike', it is likely that is has been overheated, burning away the flux.

Desoldering Methods

A soldered joint that has not been made properly will be electrically 'noisy' and unreliable and is likely to get worse in time. It may not be making any electrical connection at all, or it could work initially and then cause the equipment to fail later. It can be hard to judge the quality of a solder joint purely by appearance, because it is not possible to determine how the joint has formed on the inside.

If the guidelines are followed, there is no reason why a perfect result should not be obtained, but things do not always go to plan. A joint that is poorly formed is often called a 'dry joint'. Usually, it results from the presence of dirt or grease preventing the solder melting on to the parts properly. It is often identified by a tendency of the solder not to 'spread' but to form beads or globules instead,

perhaps partially. Similarly, if it seems to take an inordinately long time for the solder to spread, that would indicate that the parts are not all as clean as they should be.

There will undoubtedly come a time when there is a need to remove solder from a joint, either to replace a faulty component or to fix a dry joint. The usual way is to use a desoldering pump or vacuum pump, which works like a small spring-loaded bicycle pump, only in reverse. A spring-loaded plunger is released at the push of a button and the molten solder is then drawn up into the pump.

It may take one or two attempts to clean up a joint this way. If the solder is particularly awkward to remove, it may be effective to add more solder and then desolder the whole lot with the pump. The nozzle of the pump is heat-proof, but care will be required to ensure that the boards and parts are not damaged by excessive heat.

An alternative to a pump is to use desoldering braid. This product is a specially treated fine copper braid that draws molten solder up into itself where it solidifies. The best way is to use the tip of the hot iron to press a short length of braid down on to the joint to be desoldered. As the iron melts the solder, the solder will be drawn up into the braid. Extreme care should be taken to ensure that the solder is not allowed to cool with the braid adhering to the work, as there will be a risk of damaging components when attempting to pull the braid off the joint.

7

LOCOMOTIVE MAINTENANCE

Maintaining and Servicing Model Locomotives

Properly maintaining and servicing locomotives will give them a longer life, which will in turn lower the cost of ownership. They will run more smoothly and efficiently, with fewer problems and thus much less frustration.

The general areas to concentrate on are cleaning, lubrication, faults, and wear and tear. There are five main points for the maintenance of a model railway, to keep trains running well:

- Clean the wheels, not just those on the locomotive but those on the rolling stock too.
- Oil and lubricate the bearings and gears.
- Check the wheel gauges.
- Clean the track.
- Clean all pickups – usually found as wipers against the inner wheel flange.

Your maintenance regime will benefit from an understanding of all the different components of your locomotives.

COMPONENTS OF A LOCOMOTIVE

There are many different components that make up a locomotive, whether it be steam, diesel or electric. Each one will have different key parts that make up its construction, but manufacturers do try wherever possible to use similar parts across a range of locomotives, at the same time retaining the essence of the prototype.

STEAM ENGINES

1 Tender
An attached rail vehicle that holds water for the boiler and fuel such as wood, coal or oil.

2 Cab and footplate
The compartment where the driver or engineer and fireman control the locomotive and tend the steam supply and firebox. This is achieved using various devices, most of which are on the rear surface of the firebox, called the 'backhead':

- A throttle lever or regulator, which controls the amount of steam entering the cylinder.
- A reversing lever, which controls the timing of the admission of steam into the locomotive's cylinders. This is required for two purposes: to reverse the locomotive's direction (for example, when shunting); and to enable more fuel-efficient operation when the locomotive is running in a steady state.
- A train brake lever, which controls the application of brakes throughout the length

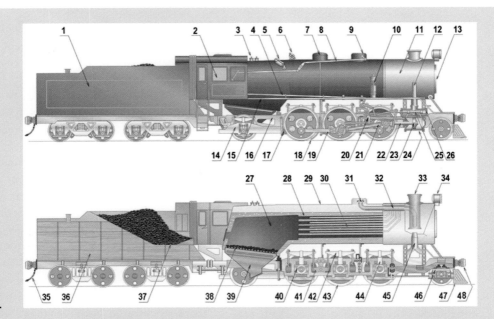

Anatomy of a steam engine.

of the train, and a locomotive brake lever, controlling brakes on the locomotive only.

- Steam pressure gauges, showing the pressure of the steam in the boiler.
- Injector valves, allowing steam to force water into the boiler when needed.
- Water gauges, for monitoring the level of water in the boiler.
- Mechanical stoker controls (fitted to larger coal-fired locomotives) or oil-feed controls for oil-fired locomotives.
- Lubricator glasses, for monitoring the flow of lubricating oil.
- A blower valve, which regulates the steam supplied to the blower.
- A whistle lever, which varies the steam supplied to the whistle.
- Blowdown (or blow-off) cocks, which allow water to be ejected from the boiler to avoid any concentration of impurities remaining after evaporation of steam.

3 Whistle

The steam-powered whistle is located on top of the boiler and used for signalling (using a varying number and length of blasts) and to give warning of the locomotive's approach.

4 Reach rod

A rod linking the reversing lever or wheel in the cab to the valve gear.

5 Safety valve

A pressure-relief valve to prevent the boiler pressure exceeding the operating limit.

6 Generator or turbo-generator

Electrical generator driven by a small steam turbine, for the headlight and other locomotive lighting.

7 Sand dome

Holds sand that is directed on to the rail in front of the driving wheels to improve traction, especially in wet or icy conditions or when there is vegetation on the line, and on steep gradients.

8 Throttle lever or regulator

Sets the opening of the throttle valve or regulator valve, which controls the amount of steam entering the cylinders, thus influencing the speed of the locomotive. It is used in conjunction with the reversing lever to start, to stop, and to control the locomotive's power output. When the regulator or throttle is closed, a vacuum valve (snifting valve) permits air to be drawn through the superheater and cylinders to allow the engine to coast freely. The throttle is not the only control that can limit the locomotive's power output. During steady-state running of most locomotives, the throttle is usually set wide open, and the power output is controlled by moving the reversing lever closer to its mid-point ('reducing the cut-off') to limit the amount of steam admitted to the cylinders.

9 Steam dome

Collects steam at the top of the boiler (well above the water level) so that it can be fed to the engine via the main steam pipe, or dry pipe, and the regulator or throttle valve.

10 Air pump, air compressor or Westinghouse pump

Powered by steam, this component compresses air for operating the train's air-brake system. The Westinghouse air-brake system is used worldwide; in Europe, the Kunze-Knorr and Oerlikon systems use the same principle as the Westinghouse. It can be a single-stage compressor or, when larger capacity is needed, a two-stage cross-compound compressor. Vacuum brakes, used in the past, do not employ a compressor. Because of their relative inefficiency, they are no longer widely used.

11 Smokebox

Receives the hot gases that have passed from the firebox through the boiler tubes and, when the throttle/regulator is open, directs them, along with steam exhausting from the cylinders, up the smokestack/chimney, sucking air through the fire bed. The smokebox may contain a cinder guard to prevent hot cinders being expelled. There are two components in the smokebox:

- The blower is a vertical pipe below the chimney petticoat pipe, with holes to blow steam upwards. It provides a draught to maintain adequate combustion – and to prevent smoke and flames from entering the cab through the firebox door – when the blast pipe is not effective enough. This may occur when a locomotive is stationary, or the throttle/regulator is closed (for example, when coasting into a station). The blower also helps to draw the fire when lighting up.
- The petticoat pipe/apron is a vertical pipe with a bell-mouth-shaped lower end extending down from the smokestack into the smokebox. Its function is to enhance and equalise draught through the boiler tubes.

12 Steam pipe

Carries steam to the cylinders.

13 Smokebox door

Hinged circular door to allow service access to the smokebox to fix air leaks and remove cinders.

14 Trailing truck or rear bogie

Wheels at the rear of the locomotive to help support that part of the locomotive and improve riding quality; see also Leading wheel.

15 Running board, tread plate, foot board or run board

Walkway around the locomotive, from the cab front, to facilitate inspection and maintenance.

16 Frame

The strong, rigid structure that carries the boiler, cab and engine unit; supported on driving wheels and leading and trailing trucks. The axles run in slots in the frames. Early American locomotives were built on frames made from steel bar. This developed in the 20th century to cast-steel frames and then, in the final decades of steam locomotive design, to a cast-steel locomotive bed – a one-piece steel casting for the entire locomotive frame, cylinders, valve chests, steam pipes, and smokebox saddle, all as a single component. British locomotives usually had plate frames made from steel plate, but some end-of-era designs included cast-steel sub-frames.

17 Brake shoe or brake block
Items made of cast iron or composite material that rub on all the driving wheel treads for braking.

18 Sand pipe
Deposits sand directly in front of the driving wheels to aid traction on steep gradients, when starting or when the rail surface is not dry and clean.

19 Coupling rods or side rods
Connect the driving wheels together.

20 Valve gear or motion
System of rods and linkages synchronising the valves with the pistons, controlling the running direction and power of the locomotive.

21 Connecting rod or main rod
Steel arm that converts the reciprocating motion of the piston into a rotary motion in the driving wheels. The connection between piston and main rod is a crosshead, which slides on a horizontal bar behind the cylinder.

22 Piston rod
Connects the piston to the crosshead.

23 Piston
Produces the motion for the locomotive from the expansion of the steam. It is driven backwards and forwards within the cylinder by steam delivered alternately, in front and behind, by the valve.

24 Cylinder
Chamber that receives steam from the steam pipe.

25 Valve
Controls the supply of steam to the cylinders. The valve gear, actuated by connection to the driving wheels, ensures that steam is delivered to the piston with precision. The different types are slide valves, piston valves or poppet valves.

26 Valve chest or steam chest
Valve chamber next to the cylinder containing passageways to distribute steam to the cylinders.

27 Firebox
Furnace chamber built into the boiler, which produces steam in surrounding water. Various combustible materials can be used as fuel; the most common are coal and oil, but in earlier times coke and/or wood were used.

28 Boiler tubes and flues
Carry hot gases from the front of the firebox to the front of the boiler, producing steam from the surrounding water. The flues are larger in diameter than the tubes because they contain superheater units.

29 Boiler
Horizontal tubular vessel, strong enough to contain high-pressure steam in a harsh working environment; closed at either end by the firebox and tube plate. Usually well filled with water but with space for steam – produced by heat from the firebox and boiler tubes – to be above the water surface.

30 Superheater tubes
Pass steam back through the boiler to dry and superheat it for greater efficiency.

31 Throttle valve or regulator valve
Controlled by the throttle lever or regulator, regulates the amount of steam delivered to the cylinders – one of two ways to vary the power of the engine (throttle governing).

32 Superheater
Provides additional heat – as much as 300°F (167°C) hotter – to steam that has been generated in the boiler by sending it back through superheater tubes located in the boiler tubes, thus increasing engine efficiency and power.

33 Smokestack, Chimney or Funnel
Vertical pipe on top of and inside the smokebox that ejects the exhaust (smoke and steam) above the locomotive.

34 Headlight

Light on the front of the smokebox to illuminate the track ahead and warn of the approach of the locomotive.

35 Brake hose

Hose for conveying force to train brakes by a differential in air pressure. Contains either high-pressure compressed air or air at lower than atmospheric pressure (vacuum), depending on whether the locomotive has an air-brake or vacuum-brake system.

36 Water compartment

Tank for water to be used by the boiler to produce steam.

37 Coal bunker

Compartment for storage of fuel before it is directed to the firebox. When the fuel is coal (or, in the distant past, coke or wood), the fireman shovels it manually through the firebox door. In larger locomotives, a mechanical stoker is used. When the fuel is oil, it is sprayed into the firebox from a sealed tank.

38 Grate

Supports the burning fuel while allowing the products of combustion – ash and small clinker – to drop through.

39 Ashpan hopper

Collects the ash from the fire.

40 Journal box or axle box

Housing for the bearing on the axle of a wheel.

41 Equalising beams, equalising levers or equalising bars

Part of the locomotive's suspension system. Connected to leaf springs, they are free to pivot about their centre, which is fixed to the frame. Their function is to even out weight carried on adjacent axles, which is especially necessary on uneven or poorly laid tracks.

42 Leaf springs

Main suspension springs for the locomotive. Each driving wheel supports its share of the locomotive's weight via leaf springs that connect the axle's journal box or axle box to the frame.

43 Driving wheels, drivers or coupled wheels

Wheels coupled to the main/side rods, through which the power developed in the cylinders is transformed into tractive power at the rails. The weight of bearings and coupling rods on the driving wheels is counterbalanced with cast-in weights to reduce 'hammering' on the track when the locomotive is under way.

44 Saddle or pedestal

Connects a leaf spring to a journal box or axle box on a wheel.

45 Blast pipe or exhaust pipe

Directs exhaust steam up the chimney or smokestack, creating a draught that draws hot gases through the firebox and along the boiler tubes.

46 Pony truck, leading bogie, pilot truck, lead truck or Bissel truck

Wheels at the front of the locomotive to guide the front driving wheels around curves, minimising yawing at higher speeds and thus reducing the risk of derailment. The truck has some side motion and is equalised to the driving wheels. It is called a pony truck or Bissel (sometimes spelt Bissell) truck when there are two wheels, while the other terms are used when there are four wheels.

47 Debris guard, snow plough, pilot or cowcatcher

A shield made from bars, cast steel or sheet steel to prevent an object on the track going under the locomotive and possibly derailing the train.

48 Coupler or coupling

Device at the front and rear of the locomotive for connecting locomotives and rolling stock.

NON-STEAM LOCOMOTIVES

Non-steam locomotives are somewhat less complicated than steam engines. The different traction types include diesel-mechanical and diesel-electric. Diesel-mechanical uses a mechanical transmission similar to that employed in most road

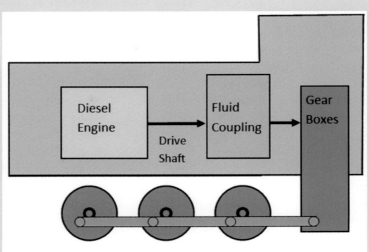

Anatomy of a diesel mechanical locomotive.

SCHEMATIC DIAGRAM OF A DIESEL MECHANICAL LOCOMOTIVE

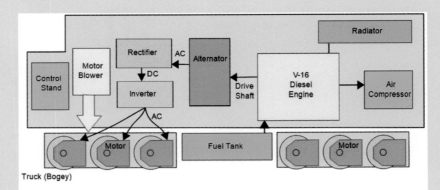

Anatomy of a diesel-electric locomotive.

vehicles. This type of transmission is generally limited to low-powered, low-speed shunting (switching) of locomotives, lightweight multiple units and self-propelled railcars.

In diesel-electric types, the diesel engine drives either an electrical DC generator or an electrical AC alternator-rectifier, the output of which provides power to the traction motors that drive the locomotive. There is no mechanical connection between the diesel engine and the wheels.

There are also diesel-hydraulic versions, electric, battery-powered, hybrid, multi-powered and hand-pump trolleys.

Tools for Maintenance

A basic tool kit for maintenance purposes can be put together simply and at a reasonable cost, although buying good-quality (and therefore slightly more expensive) tools will pay back in the long run. There is no real need for specialist tools, but there are some essential items:

- selection of jeweller's screwdrivers;
- good-quality cutters;
- small-jawed pliers;
- fine tweezers;
- fibreglass brush.

Screwdrivers

When choosing screwdrivers, ensure that the tip is correct for the type of screw. If unsure, test-fit the tip into the screw. It should fit snugly and not tend to 'cam out'. There are a few different types:

- slotted: probably the most common type, but prone to slippage and wear;
- Phillips: come in sizes 000, 00, 0, 1, 2, 3 and 4 (largest);
- Pozidrive: improved version of Phillips, with less 'cam-out'. (Note: Phillips and Pozidrive are not interchangeable, as the interior angles are different.)
- Torx: anti-tamper; not often used in model applications;
- Torx plus: improved version to allow greater torque;
- Allan-style: often used in group or set screws.

The features of screwdrivers vary between the different manufacturers. Generally, when choosing items to assist in the maintenance regime it is advisable to look for a hardened steel tip, a long thin shaft, a comfortable grip, and a swivel top.

 Slotted

 Phillips

 Pozidriv

 Torx

 Torx Plus

 Allan Style

Screwdriver tip styles.

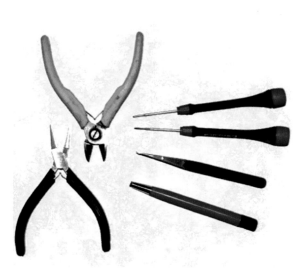

Essential tools.

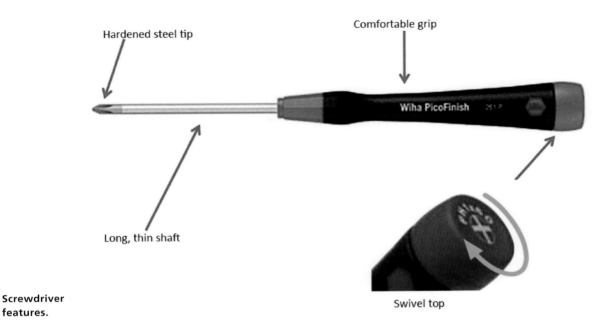

Hardened steel tip

Comfortable grip

Long, thin shaft

Screwdriver features.

Swivel top

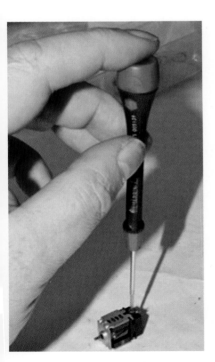

Using a good screwdriver.

Low-cost precision screwdriver.

Cutters and Pliers

Ideally, go for the better-quality cutters and pliers, which have a fulcrum that gives adjustability. However, riveted types will also give good service under normal usage.

Pliers tend to be used for adjusting and straightening rather than gripping, and flat jaws are preferable to serrated ones.

Adding a good pair of fine cutters to the tool kit is highly advisable. These can be reserved for cutting copper wire only, while lower-quality cutters can be used for other applications.

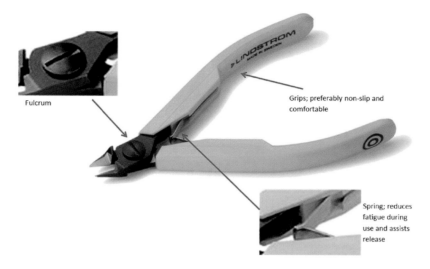

Fulcrum

Grips; preferably non-slip and comfortable

Spring; reduces fatigue during use and assists release

Cutters and pliers.

Optional Tools

- Magnetiser/de-magnetiser: magnetises the tip, which helps to pick up tiny screws when needed.
- Ultrasonic bath: for cleaning oil off inaccessible places on all parts.
- Craft knife set: for use with repairs and rebuilds; a sharp knife will always be useful to have to hand.

- Soldering iron: this is an essential tool for the general toolbox, but optional for maintenance purposes.
- Rolling road: ideal for running in, as it permits close observation on the bench. There are many versions available, and it is simply a matter of selecting the one that will be most comfortable.

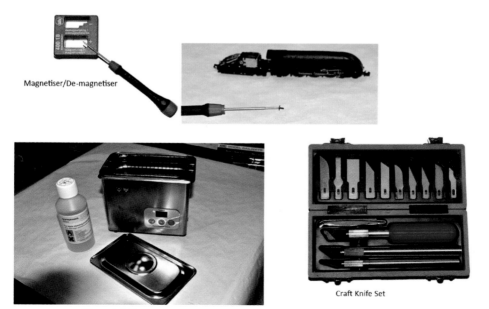

Magnetiser/De-magnetiser

Ultrasonic Bath

Craft Knife Set

Optional tools.

Soldering iron.

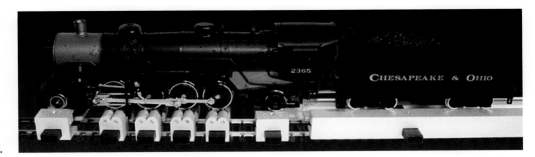

Rolling road.

Routine Maintenance

This can be split into two main categories: cleaning and lubricating. Alongside the routine maintenance programme, there will also be faults to investigate and general wear and tear to keep an eye on.

Cleaning

When cleaning, it is essential to check not only the wheels on both locomotives and rolling stock but also the pickups and track. Cleaning all the wheels ensures that they leave fewer deposits on the track. If the wheels and pickups are clean, the DCC signal will encounter less resistance as it sends commands to the decoder.

Part of the cleaning regime should also include cleaning of the track, to remove any residue or oxidisation on the surface.

Wheel Cleaning

There are different methods of wheel cleaning, from manual to drive-through. It is possible to buy a commercial device that assists in spinning the wheels while the grime is removed.

It is essential not to forget the non-driven wheels. Similar methods to those used with locomotives can be used, however the wheels must be spun manually, or with external power.

Pickup Cleaning

The cleaning regime should also include the pickups. These tend to be forgotten but they are an integral part of good communication between track and decoder, and at the same time provide an ideal place for fluff and other debris to collect. It is essential, however, to be very careful when handling them, as they can easily be bent

Wheel-cleaning methods.

Contact cleaning strips.

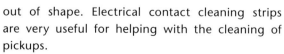

Wiper pickups.

out of shape. Electrical contact cleaning strips are very useful for helping with the cleaning of pickups.

There are a multitude of pickup types, all with their own pros and cons. The following are some of the most common:

- Wipers: mounted on the chassis and rubbing on the inside edge or running surface of the wheel. They are simple and effective and are easy to replace (if correctly designed). On the negative side, the tension of the pickups can decrease, pickups can be damaged by debris from the track

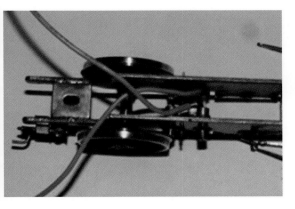

Plunger-type pickups.

Pin-point axle pickups.

Slip-type pickups.

or when cleaning, and the pickups can add to the rolling resistance of the wheels.

- Plunger-type: a variation on wipers and often fitted to large-scale models. The graphite types have low mechanical resistance, but they can have a tendency to jam, which requires maintenance. Replacement can also be tricky.
- Slip-type/journal bearings: commonly used on split-chassis locomotives, this type is less prone to

dirt than others. On the negative side, the pickups are difficult to clean and special care is needed when oil is applied as too much can act as an insulator.
- Pin-point axle: popular in smaller scales, these have the lowest rolling resistance and are simple to clean. On the other hand, they are prone to 'fluffing' (gathering fluff, dust, and so on, from the track), and the plating can wear over time.

Track Cleaning

If there is a track-cleaning wagon to hand, it should be used regularly. Some have a vacuum cleaner included, which will help to pick up all those loose bits that may be on the track – scatter, screws, wire fragments, dust, and so on. Otherwise, the track can be maintained by regular use of a track rubber.

Cleaning Fluids

When selecting a cleaning fluid, it is essential to have something that will do a good job yet not cause any damage. Most cleaning fluids are good for wheels, track and general use. All to some degree can affect plastics or paint, especially if they are left to 'pool' on a plastic or painted surface. The area must always be wiped dry after cleaning.

It is best to avoid those fluids that will leave a residue – for example, blue methylated spirits – as this will affect the running of the locomotives, especially on DCC.

There are a few types of cleaning fluid, with slightly different properties:

- A petroleum-based fluid (for example, Goo-Gone) is great for removing greasy/sticky residue. It will evaporate eventually but it will take somewhat longer than a spirit-based fluid.
- Spirit-based fluids such as isopropyl alcohol are good all-rounders and will evaporate quickly. They are usually 99.9% pure alcohol, so are highly flammable and have an odour. Should be used only in a well-ventilated room.
- Organic cleaner (for example, Locolube Track Cleaner from DCC Supplies) has a low odour and low flammability, and evaporates almost as quickly as isopropanol. Ideal for using where there is limited ventilation in the model room.

Lubrication

Lubrication is something that needs to be done at regular intervals, but perhaps only twice a year. This is not a weekly or monthly task, unless of course the lay-out is run every day for hours on end. (Unsurprisingly, an exhibition layout will require more regular cleaning and lubrication than a home layout.) Each manufacturer will have their own recommendations on frequency, but the best advice is to carry out lubrication only when needed and always in moderation.

Only a very light synthetic oil should be used, and it should be applied sparingly. Thick or mineral oils such as 3-in-1 or similar are not appropriate, and in most instances there is no need to use grease. Light synthetic greases and thicker oils may be used on larger-scale models and metal gears, but on smaller-scale models and those with plastic gears, the thinner the oil is, the better. Metal on metal and metal on plastic movements will benefit from a small amount of oil.

Over-oiling can be worse than under-oiling, leading to running issues and burning motors. In some cases, it may mix with dust and debris, forming

Types of lubrication.

a paste, which is worse than no lubrication at all! One drop is usually sufficient. Also, if too much oil is used and there are traction tyres fitted to any of the wheelsets, the rubber of the tyres will absorb the oil, which will cause them to enlarge and thus render them useless.

Manufacturers usually recommend an initial oiling on a brand-new locomotive. This is mainly to assist in thinning grease that may have thickened during transit. In this instance, a little lubrication will ease the mechanism and reduce noise.

There are a number of points to keep in mind when lubricating a locomotive as part of the regular maintenance regime:

- Oil the correct parts. Do not lubricate any bearings used in the electric path and clean off any oil, dirt or grease that might be present. Where bearings are used in the path, the manufacturer will have used self-lubricating materials that do not need any further lubrication.
- Apply sparingly. Use a dispensing tip (not usually the nozzle of the bottle, but a precision tip) to place the smallest possible drop of oil on each bearing (including motor bearings), pivots and sliders. Remove any excess with a piece of kitchen roll or similar.
- On a steam locomotive, the motion should also be lubricated.
- Diesels have relatively simple oiling requirements. Plastic-on-plastic gears require very little or no lubrication.
- Over-oiling can cause all sorts of problems, so keep the amounts of oil very small.

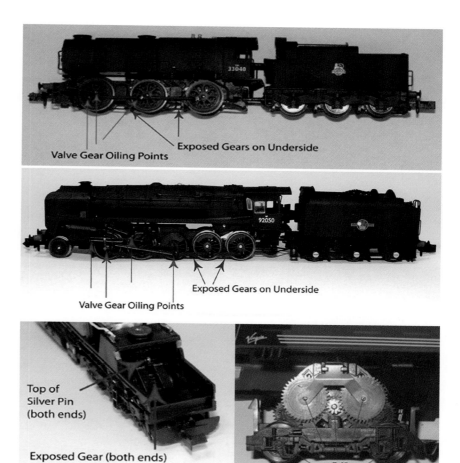

Valve Gear Oiling Points

Exposed Gears on Underside

Exposed Gears on Underside

Valve Gear Oiling Points

Top of
Silver Pin
(both ends)

Exposed Gear (both ends)

Oil

Exposed gears

Exposed gears

Common oiling points. Beware of lubricants as they can cause issues. If any parts become contaminated, remove and clean with isopropanol (isopropyl alcohol) or similar.

Repairs, Stripping and Rebuilding

There can be various reasons why a repair is necessary. It may be because of wear and tear on pickups, traction tyres and mechanicals. This is especially likely with exhibition stock that is run for extensive periods of time. Motor brushes will often require replacement after some years of service and magnets can also lose power on older stock.

'Can' motors will usually need exchanging at some point. Although they are usually designed for long service they cannot normally be repaired.

Traction Tyres

Tyres should be a tight fit and sit evenly on the rim of the wheel. It is important to clean the recess in the rim before refitting a tyre. When fitting traction tyres, do not use any lubricant. Ease the tyre in place with

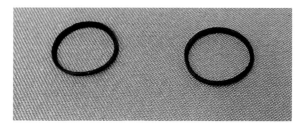

Traction tyres.

Gears with a tooth missing.

a blunt screwdriver or pointed swab. After fitting, examine it for snugness and evenness. If necessary, lift and refit. If there is any minor unevenness during a test run, it should gradually level out, so let the locomotive run for a while.

Gears

Plastic gears can crack or lose teeth, especially on older locomotives. This in turn can cause jamming or lockups. Noise can also be caused by worn teeth or by enlarged centres. It is essential to check the gear shafts for wear, as this will also cause running issues and/or noise. Plastic/metal interfaces – for example, worm gears – can be troublesome, as the plastic side can wear. In these instances, the only option is to replace the parts.

Problems with Motors

In simple terms, an electric motor works by the current causing the coil to rotate. A small motor

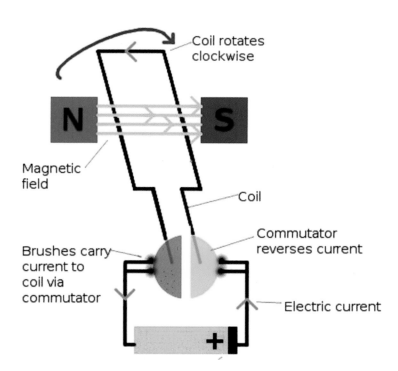

The workings of an electric motor.

Open-frame motors.

has a low power requirement. Unusually high-power usage may indicate that the brushes are worn, the commutator is dirty, there is a low winding resistance or there is a loss of magnetic power. All these issues will create additional heating, which in turn will accelerate the demise of the motor:

A motor can be tested as follows:

- On a visual check, is there any discolouration or physical damage? Can you see a dirty or worn commutator, a bent shaft or a damaged worm?

- Does the motor spin freely by hand or is it stiff and/or notchy?
- Check the winding resistance using a multimeter. Measure the resistance across the motor from one brush to another, then rotate the armature by hand and watch the meter. A good winding will read a few ohms. Open or short readings indicate a bad winding.
- Measure the current usage (amperage) of the motor. If the measurement is showing a higher current then this indicates a possible issue. Check the current usage against a similar motor in a known good model.

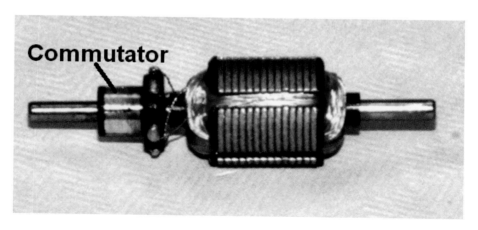

Cleaning commutators.

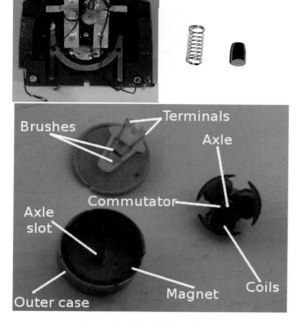

Pancake/ringfield motor.

Commutators collect lots of carbon dust. If this is mixed with oil, it will create a conductive paste! If it is not possible to gain easy access to the commutators for cleaning, use a spray.

If a motor has accessible brushes and springs, these items may need replacing.

Strip and Rebuild

At some stage, it may be necessary to strip down a locomotive to give it a deep and thorough clean. In this instance it will be essential to have a good understanding of how the locomotive was built, so that it will all go back together properly, with no spare parts left over. It is advisable to take photos of the locomotive at each stage of stripping down, to provide a reference when rebuilding.

All manufacturers have their own method of fixing a locomotive's body to the chassis. Some use clips and others use screws. The relevant information should be found in the locomotive's manual.

Can motors.

For example, to remove a Dapol tender body, the advice is to grasp the tender and chassis while allowing the locomotive to hang, then gently pull it apart from the rear of the tender. On a Hornby locomotive, the body is normally unclipped – thin plastic strips can be used instead of the fingernail method. On a Bachmann, there are usually several screws holding the body to the chassis. The trick is to find them all…

Bachmann diesel body removal.

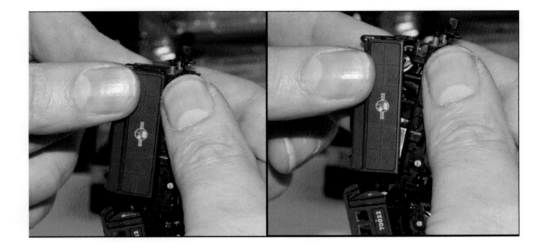

Dapol steam tender body removal.

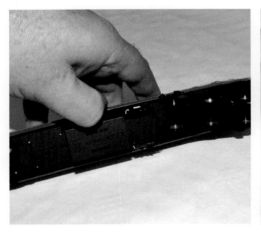

Hornby diesel body removal.

8

THE BASICS OF DCC SYSTEMS

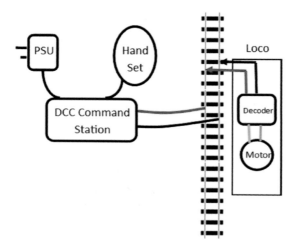

Basic DCC system.

ESU ECoS.

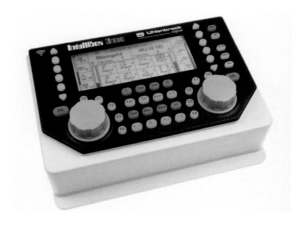

Uhlenbrock Intellibox.

Choosing a System

In principle, a basic DCC system requires just two or three items to get it up and running. The number depends on the type of system that is desired. The important factor to remember with any DCC system is that they all do the same thing – control locomotives, turnouts and accessories via digital command control – but in a different manner. There are two main types: the desktop unit and the unit with a handset. There are also systems that may be used with a smartphone or tablet.

The desktop unit is a complete command and control station with a power supply. The main manufacturers of these units include ESU, Uhlenbrock, Sig-na Trak, Digitrax, Hornby and Märklin. The modules are complete with the command station, power booster and control module.

The only other element that is needed is the power supply. Today, the majority of units come with their own power supply, so there is normally no need to worry about choosing the right one. However, if you are considering buying a US unit, it is important to be aware that it is not permitted to use a US power

Sig-na Trak ACE.

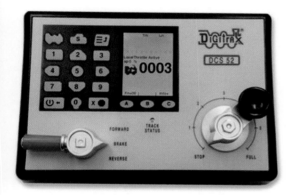

Digitrax Zephyr Xtra.

Hornby Elite.

Märklin Central Station.

The more popular systems are a combination of command station and handset, with a power supply. There are many systems on the market, from manufacturers such as NCE, Digitrax, Lenz, Gaugemaster, MRC, Bachmann, Roco and TCS.

More recently, manufacturers have been producing command stations for use with a smartphone or tablet. Some even offer voice control. These units include the Roco Z21, the MTH DCS Wi-Fi Interface, the Digikeijs DR5000 Digicentral and Games on Track.

When looking to purchase a system, the best one to choose is the one that you are most comfortable with. The best advice is to try before you buy and find someone who will be prepared to spend some time with you to explain the differences and will let you have a go.

supply in the UK. The voltage is not correct and there is no acceptable plug adapter that can be used permanently. Most manufacturers now identify their products as UK or worldwide versions, so it is possible to select the right one when browsing.

NCE Power Pro.

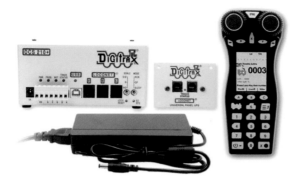

Digitrax Evolution.

Gaugemaster Prodigy Advanced.

Lenz Command Station.

Bachmann Dynamis.

MRC Prodigy Advanced.

Roco Z21 Multimaus.

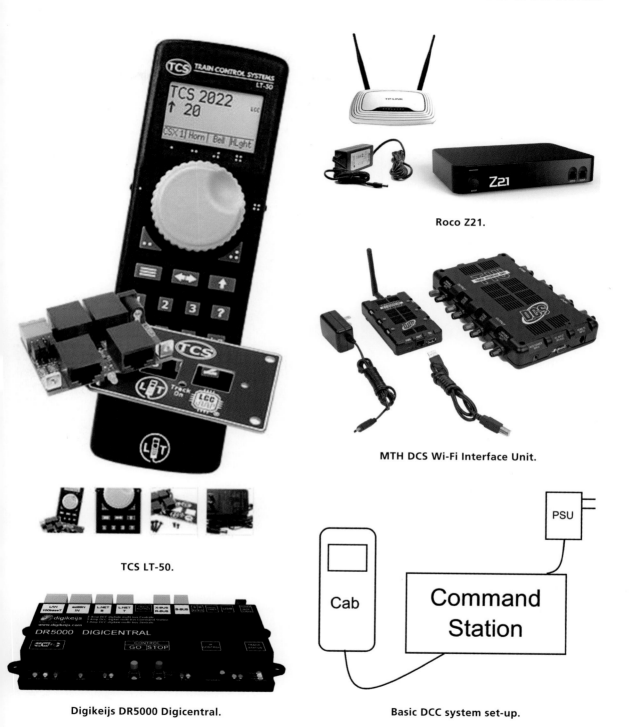

Roco Z21.

MTH DCS Wi-Fi Interface Unit.

TCS LT-50.

Digikeijs DR5000 Digicentral.

Basic DCC system set-up.

Games On Track
GT-Command.

DCC System Set-Up and Programming

Once you have chosen and acquired a system, you need to know how to set it up and get it going. First, the unit should be positioned where it will be most convenient to operate. Bear in mind that, if the system has a throttle or handset, the command station can be put out of the way and does not necessarily need to be obvious. It should, however, be easily accessible.

Depending on the system that has been selected, there should be a main-track connection and there may also be a programming-track connection. The connection for the main track needs a dropper from the track or main power bus (optional) feeding into the main-track connector, which should be marked on the system. Once this has been wired in, the system will be ready to use. Select a locomotive address to start with and off you go – in simple terms!

The programming track is used to address the locomotives and to read back CVs, as this may not be advisable or even possible on the main track. It does not necessarily need to be connected all the time, so a separate piece of track may be kept specifically for programming. This should be prepared with connecting wires and the relevant terminal block required for the system and then brought out when any programming is required. Use of the programming track is totally independent of the main track and has a much lower voltage than the track, as programming does not need a lot of current. It

is essential to ensure no other electrical accessories, such as accessory decoders, are attached.

Some people like to have the programming track as part of the whole layout. This is perfectly fine, but it must be fully isolated from the main track and have a DPDT switch to make sure that when in use there is no connection to the main track. If this is not in place, everything on the layout will be programmed and, should there be anything else attached to the track (as might be the case on a main track), this will also be programmed. Depending on the item, it may also interfere with the programming.

When programming, it is also vital to ensure that the track and wheelsets are perfectly clean, as any interference from dirt will affect the programming.

Because the current to the programming track is lower, some sound decoders may require a boost. There are modules available that can achieve this.

9

LOCOMOTIVE DECODERS – DCC AND DCC SOUND

The Role of the Decoder

There are so many different makes, models, shapes and sizes of decoder on the market, in both standard DCC and DCC sound, it can be difficult to determine which is best for your layout.

In brief, the decoder is the module that takes commands from the control system and operates the locomotive according to those commands. The clever part is that the decoder liaises with the motor to ensure the smooth operation and more prototypical running of the locomotive. It is a bit like cruise control in a car: the desired speed is set, and the engine will keep to that speed regardless of the state of the road (uphill, downhill, and so on). The control of the locomotive is the same: it will maintain the speed that has been set, whether it is going up or down an incline, or pulling wagons or carriages. The decoder controls all of this through BEMF (back electro motive force).

The other activity of a decoder is to operate functions such as lighting, beacons, catenary, doors, firebox flicker, smoke, and so on. Basically, it can control anything that can be switched on and off.

Sound decoders have the added benefit of operating sound. The main driving sounds work with the BEMF to ensure that they are coordinated as closely as possible to the movement of the locomotive. There are also other playable sounds that can be activated as desired, such as a horn, the slam of a door, a buffer clash, a squeal of wheels, and a guard's whistle, to name but a few.

Decoder Types

It is good to understand that consideration needs to be given to the gauge of the locomotives: T, Z, N, TT, OO, HO, O, 1, G, and so on. As decoders become increasingly sophisticated and tend to shrink, there are more options for the much smaller gauges. In addition, as manufacturers move towards building locomotives with DCC in mind, motors are becoming more efficient in the larger gauges. As a result, there are more options for everyone.

With the larger scales and older models, it is essential to understand the current draw of the motor, as this will dictate which decoders need to be considered. Every decoder has a constant current output and a peak output. Checking the stall current of the locomotive motor will indicate the maximum peak output the decoder should be able to go up to, so as to avoid it blowing the first time a locomotive is pulling a weight or going up an incline and using much more power.

There is an easy way to check the stall current of a locomotive. Using a multimeter attached to the track, run the locomotive, then hold it briefly so that it is struggling and quickly read the current draw off the meter. Leaving it to struggle for too long could damage the motor, so make sure everything is in place and ready to go before running the locomotive. Once you have this peak reading, it will be possible to check out the decoders.

In general, the smaller N, OO and HO gauges tend not to draw more than 2 amps, with a continuous

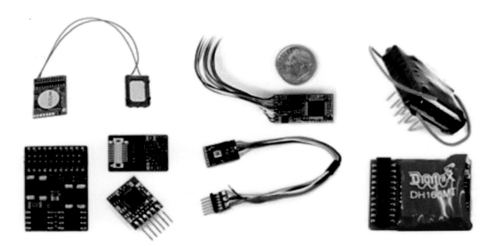

Various types of decoder.

current draw of somewhere around 0.3 amps. Some of the older OO gauge locomotives – Wren, Lima, Hornby, Bachmann, and so on – may draw more, especially if the motors are the much older versions and not 5-pole motors as most tend to be now.

Coreless Motors vs Iron-Cored ('Traditional') Motors

'Traditional' motors have a heavy iron core around which the copper armatures are wound, while the armatures of a coreless motor are wound around an external former without an iron core. The magnet of a coreless motor is placed inside the hollow armature, creating a very compact package.

Coreless types have a number of advantages:

- They can fit into a smaller space, due to their compact build.
- Overall, they are lighter than iron-core types (size for size).
- Acceleration/deceleration is faster due to lower inertia.
- They are electrically efficient – some versions at over 90% (traditional motors approximately

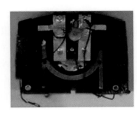

Selection of motors (top, right to left): coreless, open-frame, can; (bottom, right to left) ringfield, pancake.

50%), which makes them ideal for battery-powered devices.

- They have a high RPM.
- Low inductance (due to a lack of iron core) means a low start-up current (less arcing), so the brushes can be lighter (and less robust). They are often made from precious metals.
- Coreless motors are smooth in operation, with less balancing required than for an equivalent laminated iron-core motor.

There are also a few disadvantages to coreless motors:

- Good-quality versions are two to three times more expensive than traditional types, due to the more specialised manufacturing, components made from precious metal, and so on.
- Although they will operate at very low speeds (due to lack of cogging effect), little torque is generated, so the advantage is negated.
- Due to lower inductance, low BEMF is generated.
- They are sensitive to heat build-up due to the compact construction and smaller surface area, and a lack of robust structure in the windings (with epoxy glue in the better-quality versions or thin plastic former in cheaper types). As a result, they are much more prone to stalling/jamming.
- Due to poor heat dissipation, a smooth power supply is preferential. Also, the storage temperature needs to be monitored. As they have to be enclosed, due to the can-type design, airflow cannot be used for cooling.
- The high RPM of most types means that they require larger or more gears for reduction.
- Size for size, the torque output is lower than that of an iron-cored type.
- There is little or no consumer serviceability.

The traditional iron-cored motors offer certain advantages:

- Higher inertia means they are more resistant to rapid changes in speed.

- They have a lower RPM.
- They are very robust, with high resilience to heat build-up, so less prone to stalling and overloading.
- There is higher torque at low RPM, overall advantage over coreless types.
- Good-quality (custom) builds can excel over coreless at a lower cost per unit, due to more flexible and more widely used manufacturing methods.
- An iron-cored motor can be designed for high consumer serviceability.

Conversely, there are some disadvantages to the traditional iron-cored types:

- Quality types require individual balancing.
- Size for size, they are heavier than coreless types.
- The high level of inductance (which is due to the iron core) means that heavy-duty brushes are required. These may not last as long as the brushes in a coreless motor (which are non-replaceable).
- The RPM may be lower than the coreless types due to higher rotational mass.

On the face of it, then, coreless motors have a number of advantages over motors of traditional construction. However, as with many engineering choices, it is the application that is key. This means a careful consideration of the model railway locomotive, its operating conditions and running requirements. In some scenarios, the many strengths of the coreless types become irrelevant, and the weaknesses may be amplified, whilst the converse is true for iron-core types.

Factors to consider when making a selection might include the following:

- Ultraslow movement, with potentially heavy loads, requires start-up and slow-speed high torque.
- Sudden transients in acceleration/deceleration are un-prototypical.
- Ability to operate with a wide range of power supplies (from poor to excellent).

- Ability to work well with digital controls, especially those relying on BEMF for motor load and speed sensing (sound decoders, and so on).
- Ability to operate reliably under mechanical duress such as drivetrain issues (foreign bodies jamming gears, for example) and variations in temperature (both operational and storage).
- Space limitations in small models mean that the motor may be in a confined space (with low airflow) without much chassis contact (heatsink). The motor may physically make up much of the model's weight, so a heavy one is desirable.
- Motors are made in batches and are becoming more customisable. Motors that are not serviceable require replacement but may not be available in the future. Motors that are designed for serviceability can often be repaired, even if OEM components are unavailable. One good example is Wrenn motor brushes, which can be replaced with pencil leads!
- When making choices, it is vital to be well informed and to look for quality. This is more relevant with coreless motors, as there are many versions available that are designed for applications such as mobile phone 'wobblers', and other high-speed, low-stress environments. Due to economies of scale, these are often cheaper than custom-made iron-core types, but they may prove to be much less resilient.

In the model train industry, both coreless and iron-cored motors have a place, as a standard simply is not required. Many of the apparent strengths of the coreless versions are actually negatives when used on a railway layout, and weaknesses can be magnified by this particular operating environment. A good-quality coreless motor is ideal where there is limited space within a small model, especially if other parts can be diecast from metal to enhance traction. However, an iron-cored type will offer better low-speed performance, require less bulky gearing, operate more smoothly with DCC and be more resilient. It will increase the overall weight of the model, as well as often being consumer serviceable.

All these factors need to be considered when a designer chooses a motor and designs the remainder of the model around that choice. Thought should be given to the lifetime of the model and to serviceability far beyond that which may be expected.

Decoders

Selecting the Right Decoder

As mentioned previously, it is not always best to go for the cheapest option when selecting decoders. Lower-cost types do not always offer the full capability for operating locomotives, so it is important to check this. If you are only looking for a means to control lighting, then you can go for the low-cost option as there is no requirement for motor control.

It is not always necessary to identify what motor is in a particular locomotive as most decoders are suitable for most motors in modern-day models, without having to change anything in the CVs. However, if you do want to find out what type of locomotive you have – whether it is an older model and not even designed for DCC or whether it is DCC-ready – there are certain terms to look out for. In most instances, it should be indicated on the packaging, but this is not always the case, especially with somewhat older models. If there is nothing on the outside of the box, the information may be found in the instruction sheet with the model.

If there is no indication, the locomotive is probably analogue only and not designed for DCC. A wired-only decoder will be needed with this type and of a size that fits the space available in the locomotive. Also, for OO upwards, it will be essential to know the stall current.

If the locomotive is described as being 'DCC-compatible', this means that there are solder pads in place where the decoder wires may be soldered on. Again, you will need a wired-only decoder of a suitable size to fit the space within the locomotive.

'DCC-ready' means that there is a DCC socket within the locomotive that has a blanking plug, so that the locomotive runs on analogue quite happily.

'Sound-fitted' means that the locomotive is equipped with a DCC sound decoder.

Once you know what DCC socket is in situ, it will be possible to find the right decoder. This differs from gauge to gauge, manufacturer to manufacturer and model to model – there is no real standard. It depends on what the manufacturer had at the time of production that would fit the locomotive. Nowadays, manufacturers are paying more attention to what functions are being built into the locomotive, and the space available.

The different types of socket found in locomotives include the following:

• 6-pin NEM651: decoders come in direct fitting with pins direct on decoder or wired harness where the pins are at the end of a wiring harness.

NEM652 8-pin

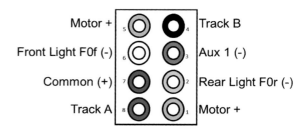

Motor + — Track B
Front Light F0f (-) — Aux 1 (-)
Common (+) — Rear Light F0r (-)
Track A — Motor +

NEM652 8-pin.

NEM651 6-pin

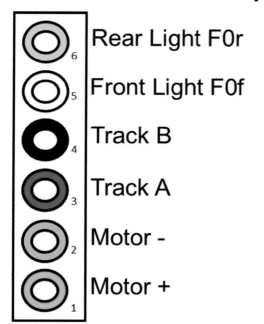

6 Rear Light F0r
5 Front Light F0f
4 Track B
3 Track A
2 Motor -
1 Motor +

NEM651 6-pin.

JST-9

1	Aux1
2	Track A
3	Motor +
4	Common +
5	Front Light F0f
6	Rear Light F0r
7	Motor -
8	Track B
9	Aux2

JST 9.

NEM660 21-pin MTC

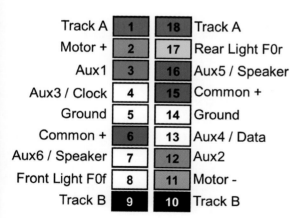

Input 1 / Aux7	1	22	Track A
Input 2 / Aux8	2	21	Track B
Aux6	3	20	Ground
Aux4	4	19	Motor +
Clock / Aux9	5	18	Motor -
Data / Aux10	6	17	Aux5
Rear Light F0r	7	16	Common (+)
Front Light F0f	8	15	Aux1
Speaker	9	14	Aux2
Speaker	10	13	Aux3
Index / Not used	11	12	Vcc

NEM660 21MTC.

NEM662 Next18

Track A	1	18	Track A
Motor +	2	17	Rear Light F0r
Aux1	3	16	Aux5 / Speaker
Aux3 / Clock	4	15	Common +
Ground	5	14	Ground
Common +	6	13	Aux4 / Data
Aux6 / Speaker	7	12	Aux2
Front Light F0f	8	11	Motor -
Track B	9	10	Track B

NEM662 Next18.

NEM658 PluX22

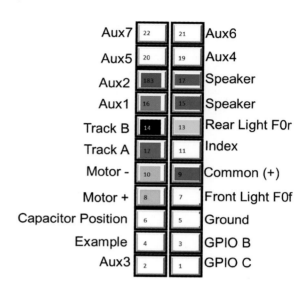

Aux7	22	21	Aux6
Aux5	20	19	Aux4
Aux2	183	17	Speaker
Aux1	16	15	Speaker
Track B	14	13	Rear Light F0r
Track A	17	11	Index
Motor -	10	9	Common (+)
Motor +	8	7	Front Light F0f
Capacitor Position	6	5	Ground
Example	4	3	GPIO B
Aux3	2	1	GPIO C

NEM658 PluX22.

- 8-pin NEM652: decoders come in direct fitting with pins direct on decoder or wired harness where the pins are at the end of a wiring harness.
- JST 9: decoders come in direct fitting with connection direct on decoder or wired harness where the connections are at the end of a wiring harness.
- 21-pin MTC NEM660: decoders come in direct fitting with pins direct on decoder or wired harness where the pins are at the end of a wiring harness.
- Next18 NEM662: typically direct-fitting, although some manufacturers have created a system where a harness can be added.
- PluX (22 or 16 Pin) NEM658: direct-fitting only.

All decoders come with a number of functions – for example, lighting options – that they can switch on and off. As an example, a decoder described as 'mobile decoder 2 function' will control the motor and two on/off functions. Typically, this sort of decoder would be used for locomotives with front and rear headlights that work directionally. The more functions the decoder has, the more possibilities there are for switching different lighting on and off. On a diesel locomotive that has both red and white lights on the front and rear of the locomotive, there is the potential to switch these on and off as required. However, this will depend on the number of functions the decoder can control:

- On a 2-function decoder, both ends are connected so that when the locomotive is running forward there will be white lights on the front and red on the rear. When the direction is reversed, there will be white lights on the rear and red on the front. The lighting function needs to be switched on by pressing the F0 button on the system after selecting the relevant locomotive.
- A 4-function decoder gives the capability to have independent control of the front and rear lighting. However, this may not have been accounted for in the locomotive itself and it may be necessary to rewire the lighting functions. This should be made clear in the user manual. In simple terms, the front and rear lighting can be split, being set up so that either the front lights only or the rear lights only can be switched on. In this way, the lighting can be more prototypical for trailing carriages or wagons, with no lights in front of them.

Steam engines do not usually have lighting fitted by the manufacturer except for firebox flicker. In this case, the only requirement would be a 2-function locomotive decoder. However, it is not an issue to use a decoder with more functions, which will not necessarily be needed or used.

To make things a little clearer, it may be best to describe the set-up of the original wired decoders. There are a total of eleven wires associated with these and this was the basis for the NMRA Standard for the manufacture and standardisation of DCC decoders.

The colours are as follows:

- Red: pickup right-hand rail (+ve side)
- Black: pickup left-hand rail (-ve side)
- Orange: motor +ve side
- Grey: motor -ve side
- White: output 1 front headlight (F0)
- Yellow: output 2 rear headlight (F0)
- Green: output 3 (AUX1 – F1)
- Violet: output 4 (AUX2 – F2), also used in some sound decoders as speaker wires
- Brown: output 5 (AUX3 – F3), also used in some sound decoders as speaker wires
- Pink: output 6 (AUX4 – F4)
- Blue: common return (+)

The mantra to follow when fitting wired decoders is easy to remember: 'Red and black to the track, orange and grey the other way'! For more on fitting decoders, *see* Chapter 10.

The original design was for a maximum of six functions but, as manufacturers started to listen to their customers and models became increasingly sophisticated, decoders began to be designed to handle more functions. As a result, there is a plethora of different fitting styles today.

In principle, once the appropriate style of decoder for the DCC-ready locomotive has been determined, the choice becomes a little simpler.

Sound Decoders

Sound decoders are not very different from standard DCC decoders. The biggest difference is that they also control sound and tend to be bigger and have a speaker attached. The principles to follow when looking for a sound decoder are the same as for any DCC decoder. However, it is advisable to know exactly what space there is in the locomotive for both the sound decoder and its speaker. If the locomotive already has provision for a speaker, it may not allow for the same size as the one that comes with the sound decoder (if it has a speaker included).

When the sound decoder does not come with a speaker, there is a question to be answered about the value of ohms that the speaker should have in order to work correctly and not blow the amplifier on the decoder. In the majority of modern-day sound decoders, either an 8- or 4-ohm speaker is required. However, there are still decoders around that require a 100-ohm speaker, and these cannot be mixed and matched. If a 100-ohm speaker is attached to a sound decoder that requires only 4 or 8 ohms, the amplifier will blow straight away. On the other hand, if the situation is the other way round, there will either be no or very little volume.

Sound decoders come both with and without specific sound projects ready loaded. Those without sounds are ready for the uploading of a created sound file, which requires a manufacturer-specific module and software. For example, if an ESU LokSound decoder is being used, an ESU LokProgrammer will be required. Similarly, if a Zimo MX or MS Sound decoder is being used, then a Zimo MXULFA module would be needed to upload a sound project.

Some manufacturers, including Digitrax, ESU and Zimo, also offer ready-made sound projects to load on to the sound decoder. They are usually free, but some files have to be paid for – this will of course save the cost of having to buy the module required for a specific sound decoder.

The source of relevant sound projects ready-loaded on to sound decoders will depend on what is being modelled – British, European, US, and so on. European sound projects can be found predominantly through ESU and Zimo, while US-based sounds are also available through ESU, as well as US companies such as Digitrax, Soundtraxx, TCS, and so on. US versions tend to be more readily available than others.

British sounds are usually handled by specific retailers who have agreements with sound recordists who have created their own projects. Some go so far as to hire a locomotive for the day so that they can record all the running and operational sounds and then create the sound project. These are then sold through the retailers as sound decoders with the sounds pre-loaded. Retailers include Coastal DCC, YouChoos, Wheeltappers DCC Sounds, Howes Models, South West Digital, Olivias Trains, Legoman Biffo and Digitrains, to name just a few. Usually, the sound decoders themselves are either ESU LokSound or Zimo MX or MS versions.

When purchasing a locomotive with factory-fitted sound, it is normal to find that the decoder has been set up to run with that specific model, in a prototypical fashion. In other words, the momentum, inertia, acceleration, deceleration and running style will all have been set by the factory. This is not to say that it cannot be changed if desired.

Some of the retailers will also have set up projects for specific makes of locomotive, but these will need to be checked out directly with them to make sure.

All sound decoders will come with a list of functions that are included in the sound project and should explain which function buttons to press in order to activate them.

Configuration Variables (CVs)

Decoders have two major benefits: not only are the locomotives being controlled, so that the operator is acting as a driver, but also every locomotive can be personalised to the running style that is preferred by the operator. Changing the characteristics of locomotive running is done by changing the relevant values of the configuration variables (CVs). The CVs are the 'pigeonholes' for storing these characteristics.

Basic CVs

There are a multitude of CVs that can be changed for both standard DCC and DCC sound decoders – the choice is down to the operator – but the main basic ones are as follows:

- CV1: short address, aka 2-digit address. Value from 1 to 127.
- CV2: start voltage (voltage applied at speed step 1). Value from 1 to 255.

NMRA S-9.2.2 - Table 1 Multi-function Decoder Configuration Variables

CV Name	CV #	Required	Default Value	Read Only	Uniform Spec	Dynamic (Volatile)	Additional Comments
Multi-function Decoders:							
Primary Address	1	M	3		Y		
Vstart	2	R					
Acceleration Rate	3	R					
Deceleration Rate	4	R					
Vhigh	5	O					
Vmid	6	O					
Manufacturer Version No.	7	M		Y			Manufacturer defined version info
Manufactured ID	8	M		Y	Y		Values assigned by NMRA
Total PWM Period	9	O					
EMF Feedback Cutout	10	O					
Packet Time-Out Value	11	R					
Power Source Conversion	12	O			Y		Values assigned by NMRA
Alternate Mode Function Status F1 - F8	13	O			Y		
Alternate Mode Function Status FL, F9 - F12	14	O			Y		
Decoder Lock	15-16	O			Y		
Extended Address	17+18	O			Y		
Consist Address	19	O			Y		
	20	-					Reserved by NMRA for future use
Consist Addr Active for F1-F8	21	O			Y		
Consist Addr Active for FL-F9-F12	22	O			Y		
Acceleration Adjustment	23	O			Y		
Deceleration Adjustment	24	O			Y		
Speed Table/Mid-range Cab Speed Step	25	O			Y		
	26						Reserved by NMRA for future use
Decoder Automatic Stopping Configuration	27	O			Y		Under re-evaluation - see details
Bi-Directional Communiaction Configuration	28	O			Y		Under re-evaluation - see details
Configuration Data #1	29	M[1]			Y		
Error Information	30	O			Y		
Index High Byte	31	O			Y		Primary index for CV257-512 00000000-00001111 reserved by NMRA for future use.
Inex Low Byte	32	O			Y		
Output Loc. FL(f), FL®, F1-F12	33-46	O			Y		
Manufacturer Unique	47-64	O					Reserved for manufacturer use
Kick Start	65	O					
Forward Trim	66	O					
Speed Table	67-94	O					
Reverse Trim	95	O					
	96-104	-					Reserved by NMRA for future use
User Identifier #1	105	O					Reserved for customer use
User Identifier #2	106	O					Reserved for customer use
	107-111	-					Reserved by NMRA for future use CV107,108: expanded Mfg. ID CV109-111, expanded CV7
Manufacturer Unique	112-256	O					Reserved for manufacturer use
Indexed Area	257-512						Indexed ares - see CV# 31, 32 Index values of 0-4095 reserved by NMRA
	513-879	-					Reserved by NMRA for future use
	880-891					Y	Reserved by NMRA for future use
Decoder Load	892	O			Y	Y	
Dynamic Flags	893	O			Y	Y	
Fuel/Coal	894	O			Y	Y	
Water	895	O			Y	Y	
SUSI Sound and Function Modules	896-1024	O			Y		See TN-9.2.3

M=Mandatory, R=Required, O=Optional

[1] If any of these features are provided, then this CV is Mandatory

Copyright 1992-2012 by National Model Railroad Association Inc.

NMRA CV list.

- CV6: mid voltage (voltage applied at 50% of maximum speed). Value from 1 to 255.
- CV5: high voltage (voltage applied at maximum speed). Value from 1 to 255.
- CVs 2, 6 and 5: the basic speed curve. Out of the box, all decoders are set to a linear speed curve, so that the locomotive will start and stop instantly, with no momentum built into the running or acceleration or deceleration rates. Value from 1 to 255.
- CV3: acceleration (the higher the value, the more time between speed steps, the slower the acceleration). Value from 1 to 255.
- CV4: deceleration (the higher the value, the more time between speed steps, the slower the deceleration). Value from 1 to 255.

SPEED STEPS

DCC runs using speed steps:

- There are 14, 28 and 128 speed steps available.
- 14 speed steps are used for G gauge and larger locomotives and not for any of the smaller-gauge locomotives.
- 28 and 128 are used usually for everything from Z gauge to Scale 1.

What does it mean? In a nutshell, rather than referring to mph or kph, it is the number of steps the locomotive's speed can go through to get from 0 speed to maximum speed. So, with 28 speed steps, there are 28 steps from 0 to maximum and in a linear format – in other words, there is the same amount of time between each speed step. This type of running could be classified as 'coarse', with a relatively small number of steps to go through to get to maximum speed.

With 128 speed steps, there are 128 steps from 0 to maximum, again in a linear format. This type of running could be classified as 'fine', with many more steps to go through to get to maximum speed.

This number is relevant when altering the acceleration and deceleration – the higher the value, the longer the time taken to accelerate or decelerate between speed steps.

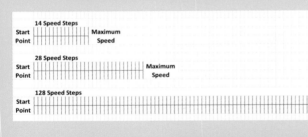

Different speed steps, from 0 to maximum speed.

Other Useful CVs

It is good to remember that there may be more useful CVs than the basic ones. These should be listed in the manual for each decoder, and may include the following:

- CV65: kick start (additional voltage applied at speed step 1). It takes a slight increase in power to get an electric motor started compared to keeping it running; this is known as stiction. If the operator wants to get a locomotive running slowly, it may be necessary to bump up the throttle speed and then back it down to the speed required. Kick start does this automatically. With DCC, there are a few ways to modify the power going to the locomotives motor (kick start and dither). Using a combination of the two allows the operator to get any locomotive started from a dead stop to nearly any speed required, and have the decoder keep it there.

- CV8: manufacturer's ID. This CV is also used in the majority of instances as the reset to factory default settings. Although the value to enter is usually 8, some manufacturers have done things differently, so it is important to read the decoder manual. A reset may be needed every now and again, as the computer can play up and benefit from a bit of a reboot.

BINARY NUMBERS

An understanding of binary is not essential, but it will help to know how the values of CVs, in particular multi-function CVs, are put together.

In everyday life, base 10 (decimal) is used, with a maximum of 9 being stored in each column. When an increase to a multiple of ten is required, there is a 'carry over' into the next column. In decimal, 99 means 9 'tens' plus 9 'ones', and 1001 means 1 'thousand' plus 1 'one'.

Converting between different bases is quite simple, but the concepts can seem a bit confusing. In the decimal (base 10) system, there are digits for the numbers zero to nine, but no single-digit numeral for ten. This is written as '10', representing '1 ten and 0 ones'. When there is a need to count to one more than nine, the ones column is zeroed out and one is added to the tens column. When there is too much in the tens column – when one more than nine

tens is required and nine ones ('99') – the tens and ones columns are zeroed out, and one is added to the ten-times-ten, or hundreds, column.

The next column is the ten-times-ten-times-ten, or thousands, column, and so it goes on, with each column being ten times larger than the one before. When writing numbers, digits are placed in each column, showing how many instances of that power of ten is needed.

The base 10 decimal system seems more 'natural' because it is the one that has been used by everyone since childhood. The reason why it was adopted? Because human beings have ten digits! Maybe in the cartoon world, where individuals have only four digits on each hand, the 'natural' base system would have been base 8, or octal.

With DCC, base 8 binary is used, which works in the same way as base 10 but in multiples of 2.

Column	7	6	5	4	3	2	1	0
Multiplier	10000000	1000000	100000	10000	1000	100	10	1
Content	0	0	0	0	0	1	1	0

Number 110 in binary (10 bit).

Bit No.	7	6	5	4	3	2	1	0
Decimal	128	64	32	16	8	4	2	1
Content	0	0	0	0	0	1	1	0

Number 110 in 8-bit binary, showing a binary value of 110 or decimal value of (1x4) + (1x2) + (0 x 2) = 6.

- CV7: firmware (software) version number. This is useful if a manufacturer has a recall or has made specific changes to decoders of a certain version.
- CV30: error information. Lenz is one of the main pioneers of using this CV for error finding, but not all manufacturers use it. Bit 0 indicates a lighting short (value 1); Bit 1 indicates motor overheating (value 2); Bit 2 indicates a motor short (value 4).

Multi-Function CVs

As the name suggests, multi-function CVs can perform a number of different functions. They rely on the fact that the CV consists of eight individual 'bits'. A 'bit' can contain a '1' or a '0' effectively acting as a switch (on or off).

CV29 is one example of a multi-function CV, with the following basic functions:

- Bit 0: locomotive direction, '0' = normal, '1'= reversed. This bit controls the locomotive's direction in DCC mode only. Directional functions, such as front and rear lights, are also reversed.

- Bit 1: speed step, '0' = 14 speed steps, '1' = 28/128 speed steps.
- Bit 2: operational mode, '0' = DCC only, '1' = both analogue and DCC running.
- Bit 3: bi-directional communication, '0' = disabled, '1'= enabled. This is only relevant for those decoders with Railcom built in.
- Bit 4: speed table, '0' = standard speed curve CVs 2, 5 and 6, '1' = extended speed curve CVs 66-95. This can be used for speed matching locomotives more accurately for regular consisting.
- Bit 5: address selection, '0' = 2-digit (short) address, '1' = 4-digit (long) address.
- Bit 6: reserved for future use.
- Bit 7: accessory decoder only.

In order to decide what to set, it is essential to know what is needed with each of the basic functions. The best advice is to have a chart to hand which lays everything out in a simple format, then switch on those functions that need to be activated and add up the values.

Say, for example, speed steps 28/128 need to be on and running in both analogue and DCC. To

Bit values.

Bit No.	1	2	3	4	5	6	7	8
Decimal	1	2	4	8	16	32	64	128
On (1) / Off (0)								

CV29 in chart format.

CV29	Running Configurations							
Description	**Direction** 0=normal 1=reverse	**Speed Step** 0= 14 1= 28/128	**Operation** 0= DCC Only 1= DC & DCC	**Railcom** 0= off 1= on	**Speed Curve** 0= Preset 1= User Defined	**Address** 0= 2 Digit 1= 4 Digit	N/A	N/A
Bit	0	1	2	3	4	5	6	7
Value	1	2	4	8	16	32	64	128
On/Off								

Value of CV29 is the total of those Bits which need to be switched on, default setting is usually 6 (2 Digit address and running on both DC & DCC)

achieve this, Bits 1 and 2 need to be switched on. Bit 1 has a value of 2 and Bit 2 has a value of 4, so there should be a value of 6 written into CV29.

Another example might be to ask what to do if a locomotive is running back to front. One answer could be to swap the motor wires round, but that would mean getting into the locomotive and desoldering then resoldering wires. A simpler solution would be to read CV29 and note the value that comes back on the system screen. If the read value is even, adding a value of 1 will swap the direction around. If it is odd, deducting a value of 1 will have the same effect. It is important to be aware that it will also change the directional lighting if that is available on the locomotive. If the locomotive lighting is working correctly but the direction is not, then the motor wires will need to be swapped round.

Function Mapping

The process of mapping decoder functions to the control system function keys is used for both physical and sound functions. Certain decoders will allow for more functions to be mapped (extended mapping); the decoder manual will give the relevant details. Zimo offers an additional different method of function mapping known as Swiss mapping. This can be quite complicated, so reading the Zimo instruction manual is compulsory.

Some functions can be dependent on others, and also dependent on commands. Basic function mapping is controlled by the NMRA. Extended function mapping, however, is almost always specific to the manufacturer and the decoder type.

Reasons for Remapping Functions

There are a few reasons for remapping functions. One is to standardise the functions of the locomotives, which is especially useful with sound decoders, so that all the common sounds are on the same buttons. Creators of sound projects all use their own form of issuing function buttons to the playable sounds and function activation. Another reason could be to simplify button presses. It may be advantageous to bring some of the functions that are in the later numbers (higher than 9), into lower numbers, particularly if they will be used more frequently.

A third reason for remapping functions may be to avoid using the momentary F2 button as found on the Gaugemaster and MRC Prodigy system. This feature is fine for sound decoders where F2 is the main whistle or horn and needs to be momentary. However, on standard DCC a momentary button on F2 means that it has to be held down in order to active the light feature on that function. This is a good example of an instance when there is a need to function-map that operation to a different function button.

NMRA Function Mapping

Function mapping is the process of changing the function buttons used to activate both on/off functions, such as lighting, as well as sound functions. The NMRA dictates the rules of function mapping which cover the following:

- All set in CVs 33–46 (although manufacturers can use reserved space for additions).
- Maps decoder outputs 1–14 to functions front light (FL(f)), rear light (FL(r)) and F1–F12.
- Contains a matrix of which controller/throttle function inputs control which decoder outputs.
- This allows the operator to select which outputs are controlled by which input commands.
- The outputs that function FL(f) controls are indicated in CV#33 and FL(r) in CV#34; F1 in CV#35, to F12 in CV#46.
- A value of '1' in each bit indicates that the function controls that output. This allows a single function to control multiple outputs or the same output to be controlled by multiple functions.
- CVs 33–37 set outputs 1–8.
- CVs 38–42 set outputs 4–11.
- CVs 43–46 set outputs 7–14.

The default is:

- F0 (f) controls decoder output 1
- F0 (r) controls decoder output 2
- F1 controls decoder output 3
- F2 controls decoder output 4
- F3 controls decoder output 5
- F4 controls decoder output 6

- F5 controls decoder output 7
- F6 controls decoder output 8
- F7 controls decoder output 9
- F8 controls decoder output 10
- F9 controls decoder output 11
- F10 controls decoder output 12
- F11 controls decoder output 13
- F12 controls decoder output 14

CV	Description	Output													
		MSB 14	13	12	11	10	9	8	7	6	5	4	3	2	1 LSB
33	Forward Headlight FL(f)														d
34	Forward Headlight FL (r)													d	
35	Function 1											d			
36	Function 2										d				
37	Function 3									d					
38	Function 4								d						
39	Function 5							d							
40	Function 6						d								
41	Function 7					d									
42	Function 8				d										
43	Function 9			d											
44	Function 10		d												
45	Function 11		d												
46	Function 12	d													

Generic function mapping table.

CV	Description	Output														Notes
		32	16	8	4	2	1	128	64	32	16	8	4	2	1	
		14	13	12	11	10	9	8	7	6	5	4	3	2	1	
33	Forward Headlight FL(f)														d	We use CV34, as it is controlled by the reverse light function.
34	Forward Headlight FL (r)												d	d		Bit 2 is set for lighting, and we set bit 3 to control the cab lights. (Decimal value for CV34 is: 2+4=6
35	Function 1											d				
36	Function 2										d					
37	Function 3									d						
38	Function 4								d							
39	Function 5							d								
40	Function 6						d									
41	Function 7					d										
42	Function 8				d											
43	Function 9			d												
44	Function 10		d													
45	Function 11		d													
46	Function 12	d														

Controlling multiple outputs with an example.

Example: we wish the cab light to switch on automatically when the loco runs in the reverse direction, but only when the headlights are on. (The Cab Light is connected to decoder output #3)

Speed Curves and Speed Matching

Regardless of whether the basic curve or the extended curve is used, the value for each CV can be set between 0 and 255. The extended curve is typically used for more prototypical running of locomotives, and is also used more with sound decoders, to match the sounds to the running.

Because of the way they are built, no two motors are exactly the same. If locomotives are to be run regularly on the layout in a double-header (consist), it will be necessary to ensure speed matching, to avoid bunching or pulling. The objective is that, using the same command station handset, each locomotive should travel the same distance in the same time. When they are all going at the same speed, this will avoid having one or more of them pushing forward (if running faster) or pulling back (if running slower).

There are many ways to achieve distance/time matching, but the most cost-effective is to set up a self-test and take measurements on a defined length of track. This is much easier if the track is an oval or round, but it can also be done on an end-to-end layout. The test is done by running the locomotives on a long section of track and timing how long each one takes to travel the defined distance.

Once the distance has been set, the locomotives are run, typically one at a time several times, at a minimum of three different speed steps. The data is gathered and then assessed so that the relevant speed curve for each locomotive can be created.

The following procedure for speed matching may take a while to complete at first, but it will become quicker with practice. The best process is to get the locomotives speed matched first at the low end (speed step 1), then at the top end (speed step 28 or 128, depending on what the system is set at), and finally in the middle. If highly similar locomotives, with the same decoders fitted, are being speed matched it can be slightly easier. If the locomotives and/or the decoders are different then it can take more time, but it is still possible and relatively simple.

Before beginning, it is essential to make sure that all the locomotives are running well, with no issues. The next step is to find out which locomotive runs the slowest at a given speed step.

At this stage, it may be a good idea to mention the availability of JMRI's DecoderPro software, which can be used to keep a log of CVs for each of the locomotives being tested for speed matching. However, although DecoderPro will make the task easier, it is not essential, as there are other methods.

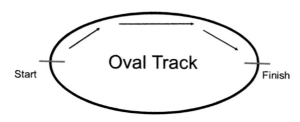

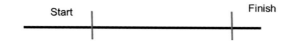

Ensure there is enough space prior to start for locomotive to come up to speed before passing Start.

Example layouts for speed matching.

Preparation

1. Make sure the testing track and all locomotive wheels are perfectly clean, to avoid any poor connectivity due to dirt or grime. Trying to carry out speed matching with poor connectivity between the locomotives and the track can quickly become a futile exercise.

2. Read the CVs from each locomotive to give the starting point for each one. If using DecoderPro, read in all the CVs to create a list per locomotive. Otherwise, read all the CVs using programming on the programming track process on the system handset and write down on a sheet of

			Loco							
			SPEED MATCHING TABLE - Standard Curve							
CV3	Acceleration	Current								
		New								
CV4	Deceleration	Current								
		New								
CV50	Motor Configuration	Current								
		New								
CV66	Forward Trim	Current								
		New								
CV95	Reverse Trim	Current								
		New								
CV2	Start Voltage Vmin	Current								
		New								
CV6	Mid Speed Vmid	Current								
		New								
CV5	Max Speed Vmax	Current								
		New								

paper all the values (*see* table above) against each locomotive.

3. There are two methods: the standard speed curve CVs 2, 5 and 6 or the extended speed curve CVs 67–94. These cannot be mixed and matched, so it is essential to know at the outset which locomotive is using which speed curve.

4. Set CVs 2, 3, 4, 5 and 6 to 0 (zero). This is to ensure that the effects of these CVs do not adversely influence any observations.

5. Turn off BEMF. The decoder manual will usually indicate the correct CV value that this should be set to, but it can vary from manufacturer to manufacturer. Sometimes, it seems to make no difference. If in doubt, turn it off.

6. All the decoders should already be set to run in 28 or 128 speed steps. Set the command station and/or handset to run in 128 speed steps. It can be done in 28, but running in 128 gives a much better result. This is not dependent on the normal running mode on the layout. If the layout is usually run in 28 speed steps, it should be returned to that setting after speed matching, and checked to ensure that everything is running as preferred, fine-tuning where necessary.

7. Get all locomotives running as well as possible at speed step 1 – starting to move smoothly and steadily – one locomotive at a time. The aim at this stage is to get them running smoothly, not to get them to run at the same speed.

8. Put all the locomotives on the layout as a double-header (consist), but do not couple them together. If there is a double track, they can be put alongside each other. Make sure all sound is turned off, or muted if there are sound decoders fitted. Run the locomotives at speed step 1 to see which one is the fastest and which is the slowest. Remember, the goal is to get the locomotives running at the same speed, with the slowest matching the fastest.

9. If DecoderPro is being used, it would be advisable at this stage to open all the entries for each locomotive (ops mode) and use the entry to make changes to the decoders. However, it is quicker and easier to use the command station handset to do the double-heading and running of the locomotives during speed matching.

10. Set the top speed of all locomotives to 180 by setting CV5 to 180 for the standard speed curve. If the extended speed curve is being used, then CVs 94 to 180 need to be set. In DecoderPro, do a 'match ends', so that the speed curve has a straight-line progression from the bottom to the top.

11. On a part of the layout that is 3–6m (10 to 20 feet) long (the longer the better), pick a section of track at least 1m long that can be seen easily

from the working vantage point and is close to the system. Mark the start and end points clearly, so that it is obvious exactly when to start the stopwatch and when to stop it as the locomotive passes.

First Step – Speed Step 1

1. First, be sure about which is the slowest and which is the fastest locomotive, then set them off one at a time at speed step 1.
2. If a locomotive moves at speed step 1, it can then be timed through the testing track. If it does not move, adjust CV2 (CV65 for extended speed curve) to the point where it will. Make a note of any CV changes.
3. After making any necessary adjustments, run each locomotive through the testing area, timing how long it takes to travel the distance.
4. The first set of results will indicate which are the slower locomotives and which are the faster ones at speed step 1.
5. To match the slowest locomotive to the fastest one, adjustments will need to be made to CV2 (the start voltage or VMin) or, if using the extended speed curve, to CV67. This is done by increasing the CV value until the slower locomotive matches the faster one. Time the locomotive repeatedly through the testing area, noting down all changes and results.
6. Continue running and adjusting until the locomotives match each other through the testing area. It does not matter which one starts first. The goal here is to get them to run at the same speed at speed step 1. It is best to repeat this process several times, until the locomotives have 'settled down' and the results are consistent.
7. When the locomotives are running at the same speed, reverse the direction and check that they run the same. If they do not run close to each other in both directions, the 'trim' CVs may need to be adjusted – CV65 and CV95 – if available.

8. Check that the speeds match by increasing to speed step 5. Time the locomotives at each different speed step to ensure they are matched as closely as possible.
9. Double-heading units while testing, but not coupling them together, can help give a visual indication of how closely the two are matched.

Speed Matching at Maximum

1. Run the locomotives at the top speed and adjust the faster one to match the slower one.
2. Keep adjusting CV5 (CV94 in extended curve) down until the faster locomotive matches the slower one. However, if the slowest locomotive is too slow for the daily running, it may be necessary to pick another locomotive that is more suitable. There is no need to change the maximum speed of the slower locomotive. Do not pay any attention to which locomotive gets to top speed first; the aim is to get the same timed running for all the locomotives.
3. Check that all locomotives run at the same maximum speed in both directions and keep making adjustments until they do. Again, it may be necessary to use the 'trims'.
4. If DecoderPro is being used with extended speed tables, redo the 'match ends' so that there is a straight-line curve on all locomotives.

Speed Matching at Mid-Speed

1. Now that the locomotives are running the same at the slowest speed and the top speed, they need to be matched somewhere in the middle. This does not need to be done at exactly the mid-point between speed step 1 and the maximum speed. Somewhere close to mid-speed will be sufficient.
2. Run the locomotives at the desired mid-speed, then change the faster one to match the slower one. VMid (CV6) will need to be adjusted (or CV80 if using the extended speed curve).
3. Check the locomotives in both directions.

4. If using DecoderPro, create one straight-line curve from the low to the middle and another one from the middle to the top, or use the sliders to create a smooth curve. Do not forget to leave the mid-point where it was set.

Verification/Validation

1. The next step is to run all the locomotives slowly up through the full speed range to check that they are running at the same speed throughout. This may be best done in stages – bottom speed steps followed by top speed steps.
2. Check that the locomotives match each other when the speed is changed. It is essential to be able to see that the locomotives increase speed and then settle down quickly for each change of speed. Stop the locomotives every so often, say, after each 10 speed steps, and check that they stop as close together as possible.
3. Should one locomotive continue to run faster or slower than the others, go back through the process of speed matching and tweak as necessary.

Momentum

1. Next, add back in the acceleration (CV3) and deceleration (CV4), referring to the notes that were made at the beginning of the exercise.
2. Run the locomotives at the varying speed ranges – slowest, mid- and top ranges. Reverse the direction at each speed range and check that the locomotives coast to a stop and reverse at the same time when the direction is changed while they are moving. A small amount of difference is not usually a problem but, if necessary, the locomotive that seems to have the 'wrong' amount of momentum can be adjusted. Remember: the higher the value, the longer the time between speed steps; in other words, the longer it takes to accelerate and to decelerate.

BEMF

Now is the time to turn BEMF back on (if it was turned off) and check to see what happens with the locomotives. This is where personal preference comes into play. When the locomotives are run with BEMF on, following speed matching, it will be interesting to assess whether there is any difference between running with BEMF on and running with it off. For some of the locomotives, it may be better to leave BEMF off. It is entirely up to the operator to decide.

Final Verification

1. For the final verification, the locomotives should not be run with the sound on. Run the double-header (consist) through the speed ranges in both forward and reverse directions. Watch them very carefully.
2. Look for bunching in the couplings. It is normal to see a small amount of clacking and stretching, however it is essential to check that the locomotives are not 'fighting' each other. This is sometimes called 'surging' or 'cogging'.
3. *Listen* to the locomotives while they are running (with no sound on). There should be no sound of any wheels spinning faster than others. If there is a problem – a result of the locomotives not having been matched properly – it will definitely be audible.
4. If there is an issue, the next step is to go back and make adjustments. If it is noticeable only in certain speed ranges, focus only on those ranges and not on all of them. It may be a good idea to switch BEMF off again to do this. Momentum should not need to be adjusted.
5. If using DecoderPro, ensure that the CV lists have been updated. Once the running of all the locomotives is satisfactory, it may be beneficial to read in all the decoder properties and save. If using a manual format, make sure that all final CV values have been noted down and marked on the respective locomotive cards.

Fleeting

Fleeting is the process of matching sets of locomotives that are always run together. It also gives an operator the ability to run any locomotive with any other, giving, say, all the same class locomotives the capability of running together. There are two approaches for doing this, neither of which is right or wrong; simply choose the one that sounds most suitable.

The first option is to divide the locomotives into 'speed classes', then work out which is the slowest of the whole batch and which is the fastest. All the locomotives can then be speed matched to either the slowest or the fastest, whichever is the preference. It is best to pick one locomotive to be used as the standard, then run through the whole speed-matching process with the main locomotive being the lead one.

If the same DCC address is to be used for the double-headed (consisted) locomotives and there are more than two, then it is best to speed match first using different address for each locomotive.

One of the easier methods of fleeting is first to match two locomotives and, once that has been done satisfactorily, double-head them with one DCC address and run them coupled. Then match the next locomotive to the double-header (consist), followed by the fourth one, fifth one, and so on, until all locomotives that will ever run in a consist have been covered. Do not forget to change the address of each locomotive to the single double-header (consist)

address once they have been speed matched. This makes life so much easier. Remember to keep checking as the process goes along, in both directions.

Function-Only Decoders

'Function-only' decoders are usually used in carriages or wagons, where there is no motor, but there are lighting applications that need to be switched on and off digitally. These decoders are used only for this purpose. They do not have any motor control.

Function-only decoders look exactly the same as locomotive decoders, although, if they are wired, they will not have the orange or grey wires, which are the ones for the motor. They should always be classified as 'function-only'. ESU refer to them by the term 'Fx' in the title. They can have a number of functions from 2 to 6, and come with the same fittings as locomotive decoders:

- wired;
- 6-pin harness;
- 8-pin harness;
- 6-pin direct;
- 8-pin direct;
- 21MTC harness;
- 21MTC direct;
- PluX;
- Next18.

Selection of function-only decoders (not to scale).

Good Products are not cheap. Cheap Products are not good!

You get what you pay for... but if there is no need for motor control, you can go for a cheaper option.

FITTING DECODERS

The Purpose of a Decoder

Before beginning to fit a decoder, it is essential to understand its purpose, whether it be a standard motor control decoder, a sound and motor control decoder or a function-only decoder.

The decoder is the 'chip' that goes in the locomotive (or tender or carriage). It may be hard-wired in place or a plug-n-play version to be put into a socket. It gives each locomotive a unique number or 'address'. The address can be either a 2-digit short address (address 1–127) or a 4-digit long address (address 128–9999). Some people use the running number as the address, or rather a version of the running number between 1–9999. (A word of advice: never use leading zeros when programming the locomotive address, as this can lead to issues with some systems not fully identifying it.) Do not be put off by the appearance of short addresses on the DCC system screen, it is usual to see, say, address '3' on the system screen appearing as '003'. The reason for this goes back to the fact that DCC uses binary coding.

The decoder detects the DCC commands from the DCC system – motor speed, function output, double-header, and so on – and implements them when 'addressed'. The decoder will drive the motor and implement momentum, BEMF, acceleration, and deceleration, depending on what has been programmed into the decoder. In addition to this, the decoder also operates any functions that are wired up in the locomotive, such as headlights, cab lights,

firebox flicker, and so on. Sound decoders have the added capacity for controlling playable sounds, which can be switched on and off via function buttons on the DCC system.

The size of a decoder is normally dependent on the current output, functions included, and whether it is standard DCC or sound.

Some decoders can be locomotive-specific, especially with US locomotives. This means that they are specially built to be a board replacement for PCBs fitting into the relevant locomotives. In this case, it is essential to purchase the correct decoder for the relevant locomotive, as they are typically not the same for different styles of locomotives. The correct decoder will allow for the specific programming of the running characteristics such as acceleration, deceleration and, optionally, the customisation of the associated speed curves for each individual locomotive.

Some decoders may have built-in specialised light and function controls, some of which will simulate lighting effects such as firebox flicker, mars lights, ditch lights, gyra lights, rotating beacons, etc.

Not all digital decoders are DCC, including some legacy systems, such as Zero 1, and proprietary systems, such as Märklin. It is therefore vital to understand what is required when making a purchase.

Note: it is possible to run a DCC-fitted locomotive on an analogue layout with hardly any difference. However, as the decoder is controlling the motor,

OUT-OF-THE-BOX DECODER SETTINGS

Decoders come with the following default factory settings:
- Short address value = 3 (2-digit)
- Long address value = 0 (4-digit)
- Address mode = Short
- Direction = Forward
- Analogue (DC) mode = On (if decoder supports it)

- Speed steps = 28 (128 steps is not a decoder setting, but a system setting)
- Speed table = Off (default straight line table)
- Advance consisting = Off (if decoder supports it)
- All special options – kick start voltage, starting voltage, momentum, light effects, BEMF – are off or at minimal settings.

the movement may not be exactly the same as it was when running on analogue without the decoder. As the decoder needs to draw power to operate, the motion may be slower than before. This becomes more relevant where DCC sound is fitted as the decoder requires around 6V before the sound can be played. This will result in the motion being noticeably slower.

Prior to Installation

Methods and Other Considerations

There are two main methods for installing decoders into locomotives, wagons or carriages. The correct method is dictated by the type of rolling-stock items:

1. Plug-in fitting: if a locomotive is DCC-ready, with a relevant NEM socket already fitted into the locomotive, wagon or carriage, the decoder can be plugged directly into position.
2. Hard-wired fitting: for locomotives that are not DCC-ready, the motor must be fully isolated from the pickups before hard-wiring the decoder into position, otherwise a short will be created.

If a locomotive is classified as 'easy to convert to DCC', it will still be necessary to hard-wire the decoder into it. This is made easier by the inclusion of solder pads that are set up for this operation. In

some instances, the pickups will already have been isolated too.

Large-scale locomotives, including O gauge, may not follow the conventional NEM fitting. Although the decoder is somewhat easier to hard-wire in a locomotive like this than on a non-DCC-ready type, there is still the need to place wires into the right terminal on the large-scale decoder board.

Whether the locomotive is DCC-ready or not, the next important step is to determine what space is available for the decoder. This will make it easier to decide which decoder to go for in terms of size. N, Z and narrow-gauge installations can be more difficult because of the limited space available for a decoder and in some cases sound fitting is nigh on impossible!

Of course, it is possible to fit decoders into tenders on steam engines or in a trailing wagon or carriage, but it is essential to be aware of the pitfalls when considering this. Any wires that need to go between the tender, wagon or carriage and the locomotive will add extra friction and drag on the train, and the lack of flexibility in the wire between the two items may cause a derailment when going round a curve or through curved points. It is also advisable to consider the size of wire used – the thinner the better, to keep friction to a minimum.

When considering sound in a locomotive, it is important to remember that this will involve the installation of a speaker, as well as the relevant decoder, thus requiring even more space. There are now very small speakers on the market, which give

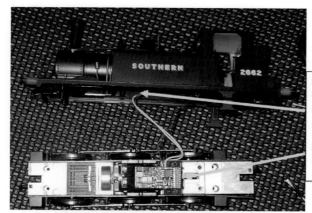

Speaker situated in the boiler –
yellow wires connect to speaker,
soldered directly to pads on main
PCB.

Decoder positioned after removing
the blanking plug

**Plug 'n' play DCC
sound decoder.**

**Hard-wired DCC
decoder.**

a phenomenal sound output for their size. It may also be possible to fit two speakers in a locomotive, which will give the option of covering more of the higher and lower tones associated with locomotive sounds. The choice will involve a locomotive-decoder-speaker match, and a knowledge of the level of ohms the speaker should have in relation to the sound decoder. When selecting a speaker, fabric cones sound better than paper or plastic ones. In addition, as speakers move air, a smaller speaker with longer cone travel (displacement) might be louder than a bigger speaker with smaller displacement. Bass sound can be improved by choosing a reflex type, which utilises air chambers and labyrinths.

Pre-Installation Measuring and Testing

First, identify whether the locomotive is DCC-ready or not. If it is DCC-ready, look at the socket that has been fitted into the locomotive to determine what style of decoder to look for. Check out the amount of space available inside the locomotive for the decoder, especially if a sound decoder and speaker are going to be installed. Plan in advance the capabilities that the decoder should have for running the locomotive on the layout, for example, load compensation (BEMF), Railcom, lighting effects, number of lighting functions, sound, and so on.

For larger locomotives or considerably older locomotives with old-style motors, it will be essential first to measure the stall current:

1. Place the locomotive without the body on a normal DC track.
2. Attach a DC current meter (ammeter) in series with one of the track feeds.
3. Apply 12V DC power to the track for N or OO/HO and some O gauges (16V for G).
4. Carefully hold the flywheel or drive shafts to stop the motor from rotating for a couple of seconds.
5. While the motor is stalled, measure the current that the unit is drawing from the power pack. To get an accurate measurement, the power to the track must remain constant at 12V.
6. Use the decoder manufacturer's recommendations to choose the appropriate decoder for the application. If the motor's stall current exceeds the decoder's maximum rating, there are sure to be problems down the road.
7. N and OO gauge locomotives with can motors draw approximately 0.3 amp. Older locos with open-frame motors or Pittman motors draw around 1.3 amps.
8. High ampage readings may indicate a tired magnet, or a problem with the mechanics or the motor.
9. Large-scale engines (O, S and G) with can motors may draw less than 2 amps; others draw more than 5!

It is always advisable to test a decoder before installation if possible. This can be done for new decoders or for those that are being moved from one locomotive to another, and can save a lot of troubleshooting later on.

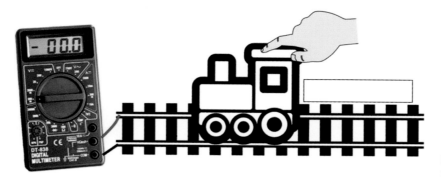

Measuring stall current of a locomotive.

Instead of using an actual motor, lights and function, it may be easier to use a test lamp to be sure that the decoder is functioning properly. Use a protection resistor to avoid any damage to the decoder caused by wiring errors. Alternatively, invest in a ready-made decoder testing module – there are various versions on the market from different manufacturers.

For first-time installers, the testing procedure will have the added benefit of helping them to become familiar with the decoder wiring prior to installation.

Hard-Wired Decoder Installation

Preparation

Once a suitable decoder has been chosen and tested, it is time to check the installation instructions, paying particular attention to the decoder wiring requirements.

Be sure to follow the manufacturer's instructions concerning light installations. Several different types of LEDs or light bulbs are used in locomotives, and some may require the use of a current setting resistor to prevent them burning out. Bulbs or LEDs that are classified for 12V will not require a resistor. Check the voltage of light bulbs in the locomotive – many small grain or rice bulbs are low voltage (around 1.5V).

When embarking on a hard-wiring project, it is important to remember the following points:

- A bad DC runner will be worse under DCC. If necessary, give the locomotive a thorough service to improve on the running prior to installation. If the running does not improve, the motor could be tired and worn out.
- Make sure that the decoder can handle the current draw of the motor.
- Converting a new non-DCC-ready locomotive will invalidate any warranty.
- Conversion may devalue a collector's item.
- Use a multimeter frequently to test for continuity, in order to identify any broken connection between motor and pickups before installation. This will also check there are no other connections to the motor that will lead to shorts when the decoder is wired in.

Step-by-Step Procedure

A decoder installed into a locomotive replaces the electrical circuit between the motor and the pickups. Because of this, prior to installation the motor must be fully isolated from the pickups. This is where a multimeter with continuity buzzer is very useful.

DECODER SPAGHETTI

Whether decoders are to be plugged in or hard-wired into the locomotive, it is advisable to know the purpose of each wire. When hard wiring, the colour of the wires will identify where the wire should be soldered into the locomotive.

Decoders follow the NMRA DCC standard wiring colours:

- RED from right-hand rail power pickup (or centre rail, outside third rail, traction/overhead wire)
- ORANGE to motor brush (+)
- BLACK from left-hand rail power pickup
- GREY from interface to motor brush (-)
- WHITE front headlight(s) power sink (-)
- YELLOW rear headlight(s) power sink (-)
- BLUE common (+) power source (all functions)
- GREEN Output 3 power sink (-)
- VIOLET Output 4 power sink (-)
- BROWN often speaker connections.

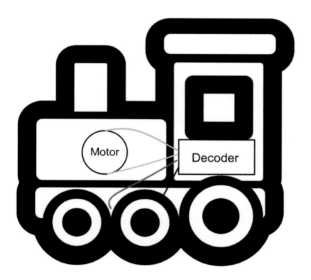

Basic connection between motor and pickups through a decoder.

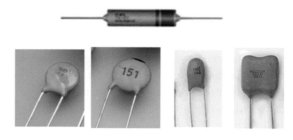

Capacitors come in all shapes and sizes.

Red & Black Wires to the pickups

Orange & Grey Wires to the motor

Basic decoder fitting.

The first step is to ensure that the locomotive runs smoothly on DC, then take the body of the locomotive off, referring to the manufacturer's instructions. Sometimes, there will be screws, while others will have a clip-on body. Be aware that screws can be hidden under bogies or coupling pockets.

With the body removed, the inside of the locomotive is visible, and it will be possible to identify the best position for the decoder. Check that there is enough space and headroom for the decoder.

If space is limited when installing into N gauge, it may be possible to remove the capacitors to free up some more. They are no longer really needed due to the fact that not many operators use AM radio any more. Capacitors are also found in other gauges of locomotives, and again are normally not required. However, unless the running of the locomotive is affected, it is best to leave them in situ. If running is not as expected, carefully snip one leg of the capacitor and see if that makes any difference. If it does, the capacitor(s) can be removed.

If there is no lighting in the locomotive, the wires for lighting will not be required. A decision will then have to be made about keeping them just in case they may be needed at a later date or removing

them totally. If the wires are to be kept, it is advisable to put a little heat shrink on the end of each one, to prevent the coating of the wires shrinking back over time and exposing the very fine wires inside. This could cause shorts later.

Next, check the locomotive again, to ensure that there is no connection between the motor and the pickups. It is now safe to take up a soldering iron and attach the wires to the relevant positions in the locomotive.

Examples of Decoder Installations

To give an idea of the pitfalls of fitting decoders, it is worth looking at examples of various installations carried out over the years. These installations may differ on newer models due to the continuous improvements carried out by the manufacturers. However, the principles are the same.

Bachmann Split Chassis (Hard-Wired Decoder)

All older Bachmann steam locomotives use a split-chassis design. One half is directly connected to the left-hand wheels, and one to the right. The can-style motor contacts use a pair of springs to connect to the appropriate half of the chassis.

The decoder installation involves the removal of the springs and the soldering of a wire to each of the motor tags. Pickups are made to an appropriate connection on each half of the chassis. In practice, this means dismantling the whole locomotive as the motor tags are inaccessible. In a case like this, it is advisable to take photographs as the disassembly takes place, to provide a reference to follow when putting everything back together.

The solders' connections are insulated with heat-shrink sleeve to prevent stray wire strands causing issues later.

In this example, the Bachmann Manor's chassis required machining to create space for the N scale decoder that was used. It was mounted using double-sided tape. It is not advisable to use electrical tape or anything similar to Blu Tack, because these items use oils and, when the motor and decoders get warm, the oils will heat up and start to leach out into the locomotives. This in turn can affect the

Bachmann split-chassis decoder installation.

motors, gearing, and so on, which may result in the locomotive ceasing to run. At that point, it will have to be stripped down and fully cleaned, hopefully removing the offensive oils.

Brass Steam Locomotive (Hard-Wired Decoder)

The decoder leads on this brass steam locomotive were connected as follows:

- Red to the engine frame (right);
- Orange to the motor lead (right);
- Black to the draw bar (left);
- Grey to the motor lead (left).

The decoder was then mounted with double-sided foam tape to the brass motor mounting bracket underneath the drive shaft.

Bachmann 0-6-0 (Hard-Wired Decoder)

This is a really simple installation. Originally the pickup wires appear at the front of the chassis and are connected via a small circuit board. The board is removed by snipping the wires, and the cast mounts are cut or filed off.

Place the decoder in the right position, fix it in place using double-sided foam tape and measure the wires. Solder the red and black wires to the pickup points, using heat-shrink sleeve to insulate the joints at the front to the pickups. next, solder the motor wires to the lugs on the motor can.

It is advisable to leave the unused function wires in place (suitably insulated and tucked out of the way), to make the later addition of lighting and other goodies easier.

Hornby Tender Drive (Hard-Wired Decoder)

This is a simple wiring installation with no lighting, but it is important to be aware of variable chassis

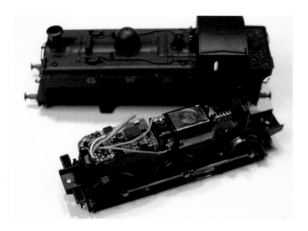

Bachmann Manor decoder installation.

Graham Farish Class 37 (Hard-Wired Decoder)

The Class 37 is similar to many Graham Farish diesel split-chassis designs, with a direct connection to each half of the chassis from the appropriate pickup. The motor is directly connected to the chassis by means of a push-fit contact. In this installation a slot was cut into the chassis where the brass tag goes, so that the motor wires could be soldered direct to the motor tags.

These chassis are made to fit many locomotives. The bogie mounts have a five-notch system. Remember to note the notch that is in use *before* the chassis is split – take photographs for reference. The locating pin is staggered to give half-notch adjustments. Again, make a note after splitting.

With the motor now isolated from the chassis sides, the motor wires are soldered direct to the brush-holder tags on the motor. The pickups are connected to the decoder by means of the chassis joining screws. The spare function wires are insulated and tucked away.

designs. Some can be metal, which will require insulation from pickups to motor.

Always check continuity to the wheels (pickups) with a multimeter and insulate and tuck unused wires out of the way.

With older models, it is a good idea to add extra pickups, as this will be beneficial to improved running.

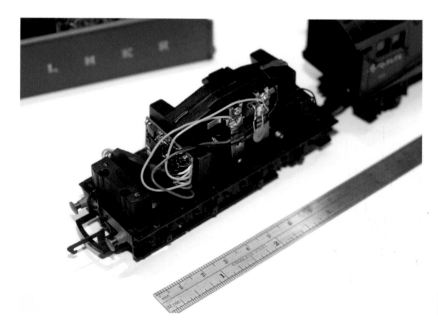

Hornby Tender Drive decoder installation.

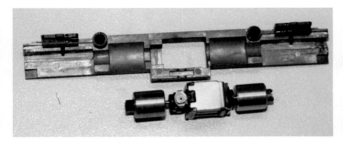

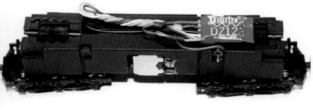

Graham Farish Class 37 decoder installation.

Some locomotives require the inside of the body-shell to be modified to make space for the decoder and wires. In this case, the decoder is fixed in place by a bit of double-sided foam tape.

Graham Farish Steam Conversions

Many older styles and in some instances newer Graham Farish Steam locomotive chassis are similar in that they use a laminated pickup design. The right-hand rail track current is passed to the motor by direct pickup contact with the chassis, the left-hand rail and via the long screw, through an insulated tunnel in the chassis to a wire above. This creates a potential problem isolating the right rail current from the motor, as the chassis itself is providing continuity.

Some early diesels are also manufactured in the same manner.

A DIY solution is to insulate the right-hand pickup from the chassis by using a length of tape. The left-hand pickup is connected to the decoder via the screw on the right; this is insulated from the chassis

Graham Farish steam chassis, showing the pickups.

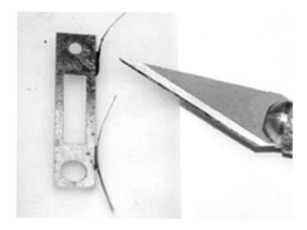

Graham Farish steam chassis – isolating pickups.

Digi-Hat brush holder.

Digi-Spring, to be used in conjunction with a Digi-Hat.

and was previously wired to the motor. This solution can also work on some diesel locomotives as well as steam locomotives.

An alternative method is to use a Digi-Hat (available from DCC Supplies), a specially designed plastic brush holder, to replace the copper brush holder in the locomotive. This insulates the brush and spring from the chassis. The brush clip legs are insulated using some Kynar heat shrink. Some diesels use a small tab instead of a clip to hold the brush spring in place. In this case, a Digi-Spring can be used.

Lilliput Triebwagen (Plug-In Style Decoder)

This locomotive was a sound project and although it appeared to have plenty of space initially, due to the seating and bonnet arrangement, there was actually very little. The customer had specified a 'large sound', so a larger speaker was also fitted.

The locomotive was DCC-ready with an 8-pin socket and the ESU LokSound decoder used was simply plugged in. The speaker was stowed above the passenger seating, and the decoder fitted under the bonnet (instead of the ideal place, which was occupied by the mouldings in the cab).

Lifelike Proto 2000 SD60 (Plug-In Style Decoder)

The light board was not needed, so it was removed by unplugging the DCC blanking plug and taking out the two screws securing it. Next, the DCC socket board was turned over so that the socket faced up. One of the screws removed from the light board was used to secure the DCC socket board into the frame.

There were two options for the lights: either replace the stock 6-volt bulbs with 12-volt bulbs or solder a 240-ohm resistor in line with the bulbs.

Finally, the decoder had a harness plugged into it via a 9-pin JST plug, and then the harness was plugged into the 8-pin NMRA DCC socket.

Lilliput Triebwagen sound installation.

Lifelike Proto 2000 SD60 plug-in decoder installation.

Athearn F3 (Board Replacement Decoder)

This installation of a TCS A4X decoder in an Athearn Genesis F3A is very simple. The decoder installation in a B-unit would be the same, except there would be no lighting to deal with.

All connections are soldered to the board, as the plastic retaining clips do not ensure solid contact.

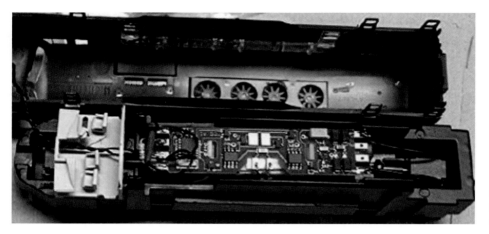

Athearn F3 board replacement decoder installation.

With Athearn locomotives, 750-ohm resistors need to be wired in series with the 1.5V bulbs.

Atlas GP7 (Board Replacement Decoder)

On this project, the original lighting board was replaced with a TCS A4X drop-in decoder.

Important note: be very careful when installing the original bulbs on the decoder. The bulb leads are not insulated and can short against components on the decoder and damage it. The bulbs must be aligned carefully so that when the body shell is installed the light pipes do not move against the bulbs and cause

a short-circuit on the decoder. A piece of electrical polyimide tape placed under the light-bulb leads can protect against short-circuits.

Micro-Trains FT (Board Replacement Decoder)

A specific board replacement decoder was used in this installation; it required the frame to be isolated from the underside of the decoder with electrical polyimide tape (Kapton).

Without an N-scale board replacement decoder, many N-scale diesels would require extensive milling in order to fit a decoder inside.

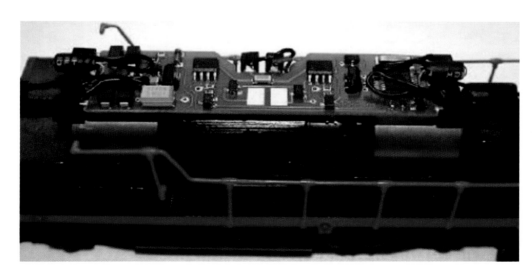

Atlas GP7 board replacement decoder installation.

HINTS AND TIPS

- Always make sure that locomotives run smoothly on DC first. If not, investigate any issues and sort them out before attempting to instal a decoder. Remember, a bad runner on DC will be worse on DCC.
- Do not forget to make sure that the decoder can handle the current draw of the motor, both in constant running and at peak.
- When mounting the decoder be sure to insulate all metal from touching it, to prevent blowing it.
- Always make sure that the motor brushes and springs are insulated from the chassis frame and pickups.
- The wheels and track need to be clean for testing and programming purposes.

- Make sure that the soldering tip on the soldering iron is perfectly clean.
- Solder all electrical connections, do not rely on clips.
- Do not rely on the drawbar on steam locomotives for pickups from the tender; it is always best to add pickup wires through to the locomotive.
- After installation, check the locomotive on the programming track for any shorts.
- Never use glue to hold things in place; they may need to be moved in the future.
- Avoid using sticky putty such as Blu Tack or electrical wiring tape. They may leach oils into the locomotive.

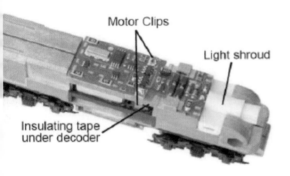

Motor Clips

Light shroud

Insulating tape under decoder

Micro-Trains FT board replacement decoder installation.

Programming Decoders

Once the installation is complete, the next step is to programme the locomotive. Programming inputs data into a memory location on the decoder, known as a CV (configuration variable). Each CV (or in some instances a bit with a CV) controls an aspect of a decoder's performance.

Write programming is the method of inputting a value into a CV. Read programming is the method of reading back the value from a CV.

It is advisable to programme only the address to begin with and then get used to the running of the locomotive before altering any running characteristics.

Programming Track

To programme the address of the DCC-fitted locomotive, a programming track will be required. The programming track is a safe place to put a locomotive after installing a new decoder – because it has limited power, it cannot blow anything up! This gives the operator the opportunity to debug the decoder installation and wiring.

Most DCC systems have a separate programming track output. If not, it is possible to create a programming track on the layout (*see* below), either as part of the layout or as a totally separate piece of track. The isolated portion of track is usually used to programme a single locomotive. If it is part of the main layout, the programming section must be isolated from the rest of the track.

It may be advisable to add a decoder protection resistor (47 ohm, 2W), especially if the DCC system

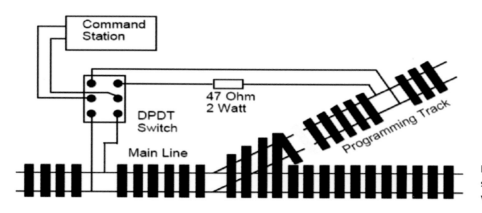

Programming track set-up for systems with no programming.

Standard Set-up

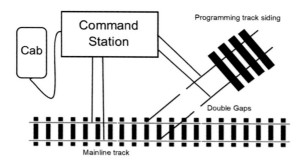

Better Set-up

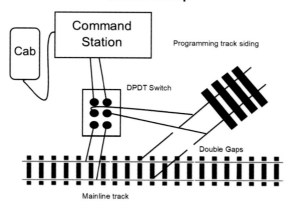

Programming track set-up – standard and better versions.

does not automatically reduce the power output to the track when programming. A double-pole double-throw (DPDT) switch will also be needed, to direct DCC from the main layout to the programming track to prevent any engines on the layout being programmed at the same time.

If the DCC system has a dedicated programming track output, there are two options for adding a programming track to the main layout:

1. Use a siding that has been isolated from the main track – this is the standard setup.
2. Add a centre off DPDT switch to allow the programming track to get normal DCC power. This allows for a quick verification of programmed settings by running the engine after programming rather than having to move it from one track to the other. This is the better setup.

Once the DCC-fitted locomotive is ready to have its address set, put the locomotive on the programming track and, using the DCC system, select programming on the programming track and set the address. As each DCC system does not necessarily programme in exactly the same way, it is best to get to grips with the basic programming process of the system in question.

The steps to programming a decoder are as follows:

1. Run the decoder-equipped locomotive from the main track on to the programming track.
2. Throw the switch to disable the rest of the layout.
3. Switch the command station to programming on the programming track mode and follow the manufacturer's instructions for programming the decoder.
4. Switch the layout back on and drive away.

The address can be set as either a short (2-digit) address or a long (4-digit) address.

Programming for Prototypical Running

There are a number of characteristics that may be considered for alteration in order to achieve more prototypical running:

- momentum – acceleration and deceleration rates (CV3, 4);
- voltage to the motor – MIN, MID, MAX (CV2, 5, 6);
- consist address (CV19);
- configuration – 14/28 speeds, brake on DC (CV29);
- speed tables (CV67-94);
- manufacturer and version (CV8, 7);
- extra lights effects (depends on manufacturer);
- load compensation (depends on manufacturer);
- PWM period/frequency (high numbers, low frequency, CV9).

Momentum effects

Momentum effects are possibly best altered using OPS mode on the main track as the changes can then be seen straight away – unless the programming track on the main layout has been set up with a centre off DPDT switch. CVs cannot be read other than via the programming track unless the decoders and DCC system have Railcom included and switched on.

Acceleration and deceleration

These are independent of each other – a train can stop faster than it starts and vice versa. Out of the box, a decoder has no momentum programmed in. The CVs required for acceleration and deceleration are CV3 and CV4 respectively. The higher the value entered, the longer it will take to accelerate or decelerate. In basic terms, the CV value sets the length of time it takes to go from one speed step to the next, whether accelerating or decelerating. This allows for more prototypical acceleration and deceleration. For example, light engines or shunters may not require an extended acceleration or deceleration, so lower values will suit better. Conversely, main-line commuter or freight trains may need longer acceleration and deceleration rates, so higher values would be better.

Speed tables

Speed tables enable the customisation of locomotive speeds vs throttle settings. In this way, they allow for locomotives from different manufacturers to all run at the same speed. This is essential for multiple-unit running or double-heading.

There are tools that exist to help set up prototype speed curves – for example, Scale Speedometer – and special programming software.

There are two types of speed curve built into most decoders: the basic speed curve with 3 points and the extended speed curve with 27 points. The CVs are: basic speed curve CV2 (min start voltage); CV6 (mid-speed – Vmid); and CV5 (maximum speed); extended speed curve CV67–CV94.

BEMF

BEMF is the 'cruise control' for DCC-fitted locomotives, giving constant motor/engine speed regardless of grade or load. The goal of this is to achieve a more realistic slowdown as load increases. When running multiple units or double-heading, the speed curves need to be optimised for successful consisting. If there is too much Back EMF, the engines will 'fight' each other, and 'buck'.

Special Light Effects

Some decoders offer special light effects, often referred to as 'FX' or 'hyperlight'. The more commonly seen types are Gyralight, Mars, Ditch, Beacon, Firebox, Rule 17, and Strobe.

FX effects have to be programmed to be activated. The decoder manual will show what CVs are relevant and give the values that need to be programmed for the effect required.

Double-Heading, Consisting or Multiple Units (Muing)

Consisting can be either decoder-based (advanced consisting) or command-station-based (basic consisting, also known as 'old style').

There are two ways of achieving decoder-based consisting. One is to set all the locomotives to be used in the consist to the same address. This causes the command station to see the consist as a single locomotive. The other way is to use the advance consisting programme, whereby the locomotives in the consist are given a temporary new address for motor control only. The locomotives then respond to a single command simultaneously – but all other decoder features are still individually controlled.

With command-station-based consisting, there are again different styles:

- With basic consisting ('old style' or 'brute force'), the command station builds a consist list of all the locomotive addresses to be used in the consist. In turn, the command station sends duplicate control commands to all the locomotives in the list. Responses can sometimes be sluggish, with small delays between locomotives. Standard or easy DCC is another term for basic consisting.
- Unified consisting is a combination of basic consisting and advance consisting at the same time.

- Universal consisting (specific to Digitrax) is unified consisting, but the command station chooses the consisting type automatically, based on the decoder capability information given to the command station in advance.

Motor Options

The Direction Bit (CV29 – Bit 0)

This bit allows for the correction of backward wiring of the decoder without the need to redo anything. The directional lighting is not affected by changing this bit. The option may not work in DC analogue mode.

Care should always be taken to wire the locomotive correctly, remembering that DC mode is sensitive to polarity.

Kick Start

The CV for kick start varies from manufacturer to manufacturer and not all decoders will have this capability. Kick start delivers a single high pulse of power to 'kick' the motor free of static friction when the first non-zero speed step is received. It needs to be adjusted for the minimum amount needed to start the motor.

Start Voltage (CV2)

Start voltage sets the minimum constant voltage needed to keep the motor running when the speed step is set to a value of 1 or higher. It is set for the slowest non-stall speed.

PWM (Pulse Width Modulation) Frequency

The CV for PWM frequency varies from manufacturer to manufacturer and not all decoders have this capability. PWM is a 'digital' method of motor speed control. It is used to adjust the pulse frequency applied to the motor. It is usually adjusted for minimum speed or motor-generated noise.

Some decoders will offer an inaudible ≥16KHz PW, for ultra quiet 'silent drive' motor operation. In this instance, the engine will run as quiet as standard pure DC analogue operation offered.

PWM does not eliminate natural motor mechanical noise.

Common Programming Problems

A number of issues may arise when programming, but the most common one is that the CVs cannot be read. This can be due to various reasons: a dirty programming track, dirty engine wheels, the decoder having the function lights left on, the locomotive having a power draw that is not under decoder control – for example, the locomotive may seem to have a short or reading the CVs may be prevented, or it may be a Hornby diesel locomotive with head code lights that are hardwired to the track.

Sometimes there may be problems with programming decoders in wagons or carriages. In these cases, as there is no motor, the command station does not receive a pulse response as it would from a locomotive. As a result, it has no way of knowing whether the command has been received and acted upon.

It is still possible to programme decoders in wagons and coaches on the programming track.

They can be treated as individual units, where the programming is carried out in exactly the same way as for a locomotive but with no feedback from the system. Once the unit is put on the main track and addressed accordingly, it will be evident whether the programming has taken or not. The other alternative is to place the unit on the programming track together with the locomotive and programme both at the same time.

Programming Considerations

Basic decoder programming will vary from DCC system to DCC system. Some systems are menu-driven, with some giving full English prompts and others using abbreviated words or two-letter codes. A basic knowledge of the techniques and terms used will be important, as will an understanding of the programming that the decoder will accept. Some of the systems use a simple series of keystrokes (with very limited options), while others will go through steps in a menu format.

Programming is the part of DCC that is closest to computer technology, so it is advisable to do some research when selecting a system, to determine how accessible the programming process is likely to be. If it looks to be too complicated, it is highly

STANDARD DECODER CVS

The NMRA Standard relating to CVs for decoders ensures that a decoder will work with any DCC system. It sets out mandatory, advisable and optional CVs to be input into the decoders. Manufacturers can choose which of the advisable and optional CVs to use in their own products, so it is important to check the manuals to see which CVs are available and thus programmable on each decoder purchased.

The standard covers the following CVs:

- Motor control – CVs 02, 03, 04, 23, 24, 65
- Mode control – CV29 (mandatory)

- Extended speed table – CVs 66-95
- Short address – CV1 (mandatory)
- Long address – CVs 17 and 18
- Advanced consist – CV19
- Decoder identification – CVs 7 and 8 (mandatory)

All DCC systems have programming capabilities to help with the rapid set-up of the most essential CVs so that locomotives can be ready to run quickly.

CV	Value/Bit	Description
1	1-127	2-Digit locomotive address (Default = 3)
2	0-255	Start Voltage
3	0-255	Acceleration Rate
4	0-255	Deceleration Rate
5	0-255	Maximum Speed
6	0-255	Mid Speed (Vmid)
7	-	Version Number (Read Only)
8	-	Manufacturer's ID (Read Only)
9	0-63	PWM Rate
17	192-231	4-digit locomotive address, higher byte
18	0-255	4-digit locomotive address, lower byte
19	1-99	Multiple Traction (Consist) Address
28	Bit	RailCom configuration
	0	1 channel 1 release for address broadcast
	1	1 channel 2 release for data and command acknowledge
29	Bit	Settings
	0	Direction of travel
		0 – normal
		1 – reverse direction
	1	Speed Steps
		0 – 14 speed steps
		1 – 28/128 speed steps
	2	Operational mode
		0 – DCC running only
		1 – DCC & DC running
	3	RailCom Communication
		0 – disabled
		1 – enabled
	4	Speed Curve Tables
		0 – factory pre-set table (CVs 2,5 & 6)
		1 – extended speed table (CVs 66-95)
	5	Decoder/Loco Address
		0 – 2-digit address (from CV1)
		1 – 4-digit address (from CV17 & CV18)
	6	Not applicable
	7	Not applicable

A list of the most commonly used CVs.

recommended to purchase or download some programming software, which will have full English statements and may also have a graphical or visual interface. This will allow for faster programming without any specialist expertise on the part of the operator. It will also simplify the programming of complex functions or features such as speed tables, special lighting effects, sound effects and Back EMF settings.

Programming Modes

There are two basic system programming modes: service mode and operations mode.

Service Mode (Programming Track)

There are four types of decoder-specific programming modes:

- Address mode – limited to two-digit address change only (CV1) and intended for toggle functions on accessory decoders, for example.
- Register mode (sometimes known as physical register mode) – does not use CV terminology. It was the original method of programming developed by Lenz, with eight registers in total of which only five can be programmed. Only very old decoders use this mode as the primary method of programming.
- Page mode – introduced the term CV as a part of its definition and is the most popular mode used today. It is quite slow at reading a single CV, so patience is required. Technically, it is register mode but with a twist. It uses a special register to point to a group of four CVs, called a 'page'; one of the four will be the CV needed at that time and it then reads or writes to the CV required. This mode was invented by Digitrax to expand the number of CVs from 8 to 1024.
- Direct mode – becoming more popular, especially as the NMRA wants it to replace page mode. This mode is much quicker at reading CVs. The

technical detail behind it is that it allows direct individual access to each CV value, so there are no complicated steps to access.

Both address mode and register mode are now virtually obsolete.

Any decoder can be programmed at any time and the mode also allows for the reprogramming of decoders where the settings may have been forgotten.

In service mode on the programming track, the DCC system can read the decoder CV values. This diagnostic capability allows for the identification of what the decoder is set to.

Operations Mode (Ops Mode)

Operations mode allows for programming on the main track. Sometimes it may be required for engines that will not programme using the programming track. This will apply mainly to sound decoders that require a little more power for programming than the programming track emits.

This mode also allows for a quick change of CVs on the main track without the necessity of going to and from the programming track to the main track. In other words, it allows the operator to programme on the fly!

Operations mode is commonly used to programme such characteristics as volume, acceleration, deceleration, and so on, while running on the main track. The locomotive address cannot be changed in this mode.

This mode does not have the capability to read back CV values unless the system and the decoder has Railcom installed and activated.

In summary, before fitting any decoder, the locomotive must be running smoothly and without any issues on analogue. The next step is to understand the locomotive in terms of space, motor current draw, and whether or not there is a DCC socket fitted, and then to determine whether DCC or DCC and sound is wanted. It will then be possible to select a suitable decoder.

11

ACCESSORY DECODERS

Accessory decoders (also called stationary decoders) are usually multi-output modules which can be used to operate lineside accessories on the layout, such as point motors, signals, crossings, lights, and so on.

These modules are normally mounted on the underside of the layout and, as with locomotive decoders, they are activated by commands from the throttle. An NMRA-compliant accessory decoder is controlled in the same way as a locomotive decoder via the throttle bus and has certain standards applied to its manufacture by the NMRA. The difference between this type of decoder and a locomotive decoder is that each output has an address rather than the decoder itself, and each output switches an item on and off only by means of a pulse of power.

The DCC systems have a special method of controlling accessory decoders – there will be an 'accessory' button to press to select the accessory decoder output address that is to be switched.

Accessory decoders come in all sorts of flavours, so it is essential to know what type is required for the accessories on the layout. These might include solenoid point motors, stall action or slow-motion point motors such as Tortoise, Fulgurex or Cobalt (to name but a few), light aspect signals, semaphore signals, servo motors, lights, turntable, and so on. Some decoders will activate multiple types of accessories whereas others will be accessory-specific, so it is vital always to read the description.

Accessory decoders tend to be designed to perform a specific function and may require a specific

Types of accessory decoders.

set-up defined by the manufacturer. The following information on addressing accessory decoders is based on the NMRA standards, but not all accessory decoders will follow these (beyond basic addressing) as they are partly optional. Additionally, few, if any, command stations support programming directly via CVs using the accessory address, so programming a decoder may depend not only on the specific decoder, but also on the command station's preferred method of programming. The decoder manufacturer should document this if it differs by command station. In short, it will be helpful to have an understanding of the programming capabilities of the DCC system and of the instructions for each accessory decoder to be programmed.

Applications

Controlling Point/Turnout Motors

The primary use of accessory decoders is for controlling point/turnout motors, and many are specifically designed with this purpose in mind. There are two common types of point motor: solenoid motors (for example, Peco Point Motors and Gaugemaster Point Motors) and stall motors, or slow-action point

Accessory decoder connections.

motors (such as Tortoise Switch Machines, Fulgurex and Cobalt). Each type of motor requires a different kind of electrical drive. The solenoid type needs a short but powerful pulse of current to snap the turnout from one direction to the other. If the current flow is not cut off afterwards, the solenoid motor will be destroyed. The stall type requires a constant but low current to drive a geared motor slowly from one side to the other. The name derives from the fact that they stall when they have reached the mechanical limit and the stall current can be maintained indefinitely to hold the switch blades in position.

Solenoid motors are best driven by a capacitor discharge unit, which uses the energy stored in a large capacitor to supply a short, powerful pulse to the electromagnet without the risk of overheating it. Some accessory decoders have this capacitor already built in. Other decoders simply allow the pulse duration to be set accordingly. Solenoid motors can be 2-wire or 3-wire. The 2-wire versions use a single coil with the polarity of the DC voltage determining the direction of throw, whereas the 3-wire versions use two coils, one for each direction.

Signalling

Another common use for accessory decoders is signalling. The types of signal typically used on layouts are either electromechanically operated semaphore arms or colour light signals. There are specific decoders on the market that will automatically set the aspects for colour light signals – for example, LDT signal decoders – and the same goes for semaphore signals (one example is the Train-Tech Semaphore Signal Decoder).

Other Uses

If there are servo motors on the layout for, say, point motors, crossing gates, and so on, then a servo accessory decoder must be used, as other accessory decoders will not activate the servo correctly.

An accessory decoder can also be used to control any other simple on/off function such as structure lighting, animation effects, crossing gates, water pumps, and so on.

Additionally, there is no real reason against using a small accessory decoder in rolling stock to control the lighting instead of the usual locomotive function-only decoder.

Outputs and Power

Some accessory decoders have only fixed-pulse outputs, which are suitable only for powering solenoid point motors. Others offer user-selectable outputs, which can provide a range from pulse through to constant power output for each set of terminals. Pulse is for solenoid point motors while constant is suitable for stall motors or signal feeds, and so on.

The number of outputs varies too. Most accessory decoders provide two separate outputs as a minimum, while some can provide up to eight. Normally each set of outputs are in groups of three, often marked as '+ C -' , where '+' is one feed, 'C' is its common connection and '-' is the opposite feed. When '+' output is turned on, the '-' output is turned off and vice versa, with 'C' common to either.

There are two types of power source suitable for accessory decoders: DCC track power, where the decoder will have specific inputs for track power, and a separate external power supply (which can be a separate power bus for powering other accessories). In the latter case, the decoder will still receive its control signals from the track. The majority of accessory decoders will accept either track power or external power and have individual terminal input connections for each type. There are two reasons why using an external power source may be preferable: to reserve the power from the DCC system purely for locomotives and to aid with troubleshooting should any issues occur on the track, such as a short on the line.

Addressing Accessory Decoders

Some command stations do not support programming of the accessory decoder address via the standard NMRA mechanism, which is the case with locomotive decoders. It is essential to read the instruction manual thoroughly and use the

Type	Description	Example Applications
Momentary Pair	Pulse output	Solenoid point motors
Constant Pair	Constant bidirectional output	Slow-action or stall point motors 2-Aspect colour light signals
Single	On/Off style output	Structure lighting

The different styles of output available with accessory decoders.

manufacturer-specific approach. One common method is to connect a turnout to an output on the decoder and then send a command from the throttle for the desired address. The decoder will then store this address to that particular output. For example, to set output address 12, attach the relevant turnout to the decoder output, then, using the throttle, select the accessory address 12 and send a 'thrown' command. At this stage, it is advisable to remove the power from the decoder. The relevant output should now operate on address 12.

There are other methods of addressing, so it is always best to read the accessory decoder manual in full to understand the address programming process for a particular decoder.

Multi-output accessory decoders offer the option of addressing the outputs sequentially – for example, 1–8 for an 8-output decoder – but first setting the address for the first output. This style of programming makes it easier to programme multiple decoders in a short time. The default address of an accessory decoder always starts at '1'. If the accessory decoder is a single-output version, this address is usually set in CV1 (or, on some older decoders, in CV513). If the decoder programming allows the operator to change the address to any value up to 63, simply write the appropriate number into CV1 (or CV513). However, if an address greater than 63 is to be used, addressing becomes a little more complex.

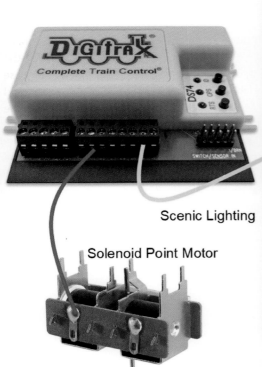

Scenic Lighting

Solenoid Point Motor

Photo used with permission from Peter Annison

As well as the locomotives, accessories alongside the track can also be controlled via DCC, using an accessory decoder that is suitable for the items that need to be switched.

GENERAL TROUBLESHOOTING

The most important piece of advice for trouble-shooting is do not panic. Think things through rationally and logically. Taking it slowly may seem a tall order at first, but it is worth it in the long run. Start at what may be the main point of the issue and work through from there, proceeding step by step. Troubleshooting is not always a quick process, and assumptions should never be made. The first questions to ask are, 'What happened?' and 'What changed just before the problem started?'. For example, to take a non-railway-related example, if a light in the house is not working, there could be a number of reasons for this: the bulb may have blown; the switch may not be working; the circuit breaker may have tripped; there may be a loose wire; or maybe the string on the fixture has not been pulled! The only way to reach a solution is to check each possibility until the issue has been sorted. And all too often it will be last check that uncovers the issue…

It is never good to make assumptions as this may lead the investigation down the wrong path. For example, if you suspect that a connection is not good, you may check all the pins on both mating parts (plug and socket) separately. If they are all fine, the assumption will be that it is not the connector, and that you can go on to look at something else. However, it would be wrong to move on until everything with the connector has been thoroughly checked and tested – for example, do all the pins mate when the two parts of the connector are pushed together?

In addition, it is important to question everything that is *not* known about an issue, and establish the facts, whether good or bad. If unknown factors are causing the issue, the issue may never be resolved. Always look at the situation broadly, try to think 'outside the box' sometimes, and avoid making assumptions. Look at the facts and make sure that you investigate all the possible causes in a calm and methodical manner.

Wiring Issues

Wiring issues may be related to either the layout or the locomotives, but they are more likely to be found on the layout. The solution may depend on whether the connections are soldered or connected via terminal blocks. If they are soldered, a multimeter can be used to check the solder joint for continuity, to ensure that the connection is still good. If continuity is not detected, then the joint needs to be remade. This is done by desoldering the current joint and then applying new solder. Adding more solder to a bad joint will not help at all.

If the connections are via terminal blocks, either as part of the main bus or connections to systems and modules, give the wires a tug to see if the connection has loosened. This will result in poor connectivity. Terminal block connections are good for joining wires or attaching wires to systems or modules, but the screws that hold the wires in place can loosen over time. This can be prevented by checking all the connections as part of a yearly or bi-yearly layout service, and tightening any screws as necessary, to make sure that the connections remain good.

If all else fails, try isolating the problem. If nothing is working and the issue cannot be resolved, isolate parts of the system and check them out one by one. Start with the DCC system and disconnect everything, including both main and programming track wiring, handsets, and power. Then plug the power back in. Does the system work? If it does, add a handset. Does it still work? Take everything a step at a time logically and work through all the connections until the problem is found:

- If there are handset panels dotted around the layout, disconnect them all from the DCC Command Station. Once a handset has been plugged directly into the command station and both have been found to be working fine, plug the adapters/faceplates back in one at a time, and check each one with a handset connected for operation. If the first one is good, move on to the next, and so on.
- Add the programming track connection. Are there any signs of a short? If not, try a locomotive on the track and read the address for instance. If this worked OK, remove the locomotive and move on to the next step.
- Add the main track connection. Is the system still good, with no sign of a short? If the answer is yes, then it seems the issue may have been resolved. However, to make doubly sure, place a locomotive on the track and try to run it. If the locomotive does not run or the system is showing a short, think back to what was last added to the track and start to isolate sections, gradually adding them back in to find out where the issue may be.
- If a short is causing the issue, and the system, connections and locomotives are not responsible, start to look at the points. Is there anything caught in the blades? Are there any points that

are set incorrectly? Is there a locomotive, wagon or carriage stopped in the points that may be causing a short due to the wheels bridging the blade and stock rail?

Troubleshooting Tools

There are a few tools that are very useful when troubleshooting:

- A multimeter with continuity buzzer is indispensable for measuring voltage, resistance and continuity.
- A coin can be used to test that the system will show when a short occurs. Check the track every metre or so to ensure that the system indicates a short or shuts down.
- Instruction manuals for everything connected to the layout, including locomotives and decoders.
- The small hand tools that were used for setting up the layout – tweezers, cutters, screwdrivers, and so on.
- Patience.

Common Troubleshooting Questions

Problem: the DCC-fitted locomotive does not move

Has the right address been entered? Check the address and try again. If it is still not moving, put the locomotive on the programming track and read back the address. Has the decoder lost its programming? Try selecting address 3 and see if the locomotive runs.

If the locomotive is a sound-fitted one, has F1 been turned on? Sound decoders are set up for a start-up procedure and the locomotive will not run until that process is complete, even if speed has been added to the locomotive.

Does the display on the DCC system indicate that the locomotive is part of a consist/multiple header? If the locomotive is not currently running in a consist, it needs to be removed from this until it is required again.

If none of this helps, try a factory reset on the decoder. Just like computers, decoders can sometimes have a 'moment' and a reset will put them right again. The normal process for resetting a decoder is setting CV8 to a value of 8, but some manufacturers have used a different method for resetting decoders, so you should always refer to the decoder manual. Remember that a reset will set the decoder address back to 3 and any momentum effects added will be returned to factory default settings. This includes sound decoders. The difference between DCC and sound decoders is that the sound decoder settings will revert to those settings created with the sound project, so there is nothing to worry about if none of the running characteristics have been changed.

Problem: the system cannot read the CV

CVs cannot be read on the main track, only on the programming track. If the system is struggling to read or change a CV, check the track and the locomotive's wheels for cleanliness. Although the layout will run well even with dirt on the track or wheels, when programming it is essential to have a very clear connectivity between the DCC system and the decoder. For this to happen, it is imperative to have a perfectly clean programming track and locomotive wheels.

If the locomotive has a sound decoder fitted and the system is still struggling to read or write CV settings, it may be that the decoder is drawing more current than the programming track output is putting on to the track. In this instance, a programming track booster may be required.

Also remember that decoders in non-motorised units will not read, due to the fact that there is no motor to pulse a response back to the system to confirm that it has received the communication. Changing CVs will still work, but there will be no

confirmation that the writing of the CV has worked. This may cause an error to show on some systems.

Problem: there is shorting or nothing is working during decoder installation

There are various questions to ask if there is an issue during installation:

- Was the motor fully isolated from the pickups prior to fitting the decoder?
- Are there any stray strands of wire touching any metal parts inside the locomotive?
- Are the pickups making good contact with the wheels?
- Are the solder joints good? Have they been tested for continuity?
- Is there anything metal touching metal within the locomotive? If necessary, use polyimide tape (Kapton tape) to isolate electronic components from metal surfaces. This includes isolating motor tabs from the chassis frame.

Problem: locomotives programmed on a different system will not run

Providing the locomotives have been correctly programmed, there is no reason why they will not work on a different system. The only time there may be a problem is when one of the systems does not conform with the NMRA standards. This is likely to occur with a much older system.

Remember, it is advisable never to use leading zeros when setting the address of a locomotive. Address programmed in as 3, 03, 003 or 0003 are all different where DCC systems are concerned. If the address has been set on one system in one way and then is called up on another system in a different way – for example, address written in as 03 and then called up with 3 – then the locomotive will not be found as the system is looking for a different address. If the address wanted is 3, programme the address as 3 and ignore the fact that the display is showing 003 or 0003.

Certain faults on a control system may also cause this problem. If all else fails, have the control system checked out.

Problems: locomotives are slower on DCC than they are on DC

The speed of the locomotive is determined by the track voltage and design of the locomotive gearing. There can be various reasons for slow running:

- Lack of proper layout wiring – best practice may not have been followed. Are the wires too small? Are the droppers very long? This could be leading to voltage droppage. Are there sufficient droppers around the layout, giving an even spread of power?
- The power supply for the system is putting out low voltage. In general, for the majority of layouts in N, OO and some O, the minimum voltage of the power supply should be 15V. This is because the track needs to have at least 12V and the DCC system will hold back a few volts. If the power supply has adjustable voltage, check that is set to 15V. Similarly, if the DCC system has the capability of setting the voltage out to the track, check that is at the correct voltage for the layout.
- The track may be dirty. Although DCC will put up with a considerable amount of dirt, there will be a point at which the dirt will affect the running. Sound locomotives require a much cleaner signal. Lack of regular maintenance can cause various issues with DCC that may be less problematic under DC. The same goes for rolling-stock wheels, especially where there is a decoder fitted in the unit.
- There is always the possibility that either the locomotive has an issue, which may be sorted by a service, or the decoder has an issue, which may be sorted by a reset.
- The decoder may have a speed cap (CV5) or momentum programmed in, which may be dictating the slower running.

Problem: the locomotive is not running right, and the lights are flashing

There are some occasions when the locomotive runs in a totally different manner from how it did when it was first set up and run round the layout. If a decoder reset does not remedy the strange running, look at the speed step to which the system has been set. If the system is showing a speed step setting of 14, this could be the issue. In general, speed step 14 is now only used for LGB or large-scale locomotives. If the system is then set to speed step 28 or 128, the locomotive should start to run properly again.

Problem: a runaway locomotive

There may be various reasons why a locomotive is running away. Decoders are typically set to run on either DCC or DC; the decoder can detect what signal it is receiving and will act accordingly. If there is something on the track interfering with the DCC signal, the decoder is seeing full power only with no DCC signal and has accepted that it is running on a DC track. This can result in a runaway!

If the DCC system is sending out spurious signals – not a true DCC square wave with both power and data packets – this will also cause the decoder to decide that it is on DC and not DCC. In both instances, a 'sticking-plaster' approach can be taken to get up and running again quickly. This involves turning off the capability of the decoder to run on DC. This is done via CV29 (see Appendix). However, this will not eliminate the real cause of the issue and it is highly advisable to troubleshoot this and try to resolve it.

The same locomotive may be selected on different throttles, both of which have different speeds set. Some systems allow for the same locomotive to be controlled by multiple handsets, whereas some will allow the moving of a locomotive from one handset to the other. In this case, it will ask whether the locomotive should be 'stolen' from the other handset.

Does the decoder allow for a setting to remember the last known speed command, for example, after a power interruption? If so, the problem could be that the locomotive has been removed from the track while running, and, when replaced on the track, has reverted to the running speed from before the interruption.

Problem: the DCC system has blown

This usually occurs because the power has been put into the wrong input and thus shorted the internal PCB. However, in very rare instances a short somewhere in the household wiring will cause a massive spike to run through the ring main and out via the shortest route – through the DCC system. In this instance, it is advisable to plug the DCC system into a 13A fused plug-through active RCD adaptor.

Problem: the locomotive runs in the wrong direction

This usually happens because the motor connections have been wired the wrong way round. If this is the case, there are two ways of remedying it. Either unsolder the wires to the motor and switch them round, resoldering them into position, or change the direction in the decoder by altering CV29. First, read back CV29. If the value returned is even, add 1 to it and enter that into CV29. If the value is odd, take off 1 and enter that into CV29.

Either of these remedies should result in the locomotive running in the right direction. Lighting should not be affected.

Problem: the locomotive runs less well than it did on DC

DCC will not improve the running of a locomotive if it already has mechanical issues or needs a good service prior to a decoder being fitted. In fact,

adding a decoder could make the running worse, especially if the issues are related to the Back EMF working with the motor. If this is not running right, the locomotive will run very badly or possibly not at all.

Problem: the locomotive does not run smoothly or does not run on speed step 1

This can be due to poor track pickup that is related to intermittent or low rail conductivity. To remedy this, check the pickups and ensure that the conductivity on the track is good all round.

It can also be caused by dirty track or dirty wheels, so the track and locomotive wheels should be cleaned regularly.

The motor may need a little kick to start it off. To do this, put the locomotive on the programming track and read CV2. Take the value and add, say, 5 to it. Programme this new value into CV2 and then test the locomotive again at speed step 1. If the locomotive now moves on speed step 1 and is smooth, it is good to go. If not, add another 5 to the new value and try again. Keep repeating this until the locomotive will move straight away on speed step 1 without stuttering.

Problem: occupancy detection false readings

Occupancy detection is used to monitor the physical progress of a train around the layout. It is also used for automation of layout control and/or signalling. One common problem with occupancy detection is a leakage current, which is a flow in unintentional current due to either a resistive or capacitive leakage path. Resistive leakage is found in the form of oils and dirt that contain conductive particles such as metal filings or graphic lubricants. The best solution for this is to clean and vacuum the track regularly, ensuring that special attention is paid to points, especially between the blades. In some instances, the oils may have penetrated any ballast around the track. If this cannot be removed by cleaning, then it is best to replace the section of ballast that may be causing an issue.

Capacitive leakage is found in the wiring of the layout. The best remedy for this is to ensure that all wiring on the layout has been subject to best practices. Do not twist wires unless there are huge continuous lengths of it. Twisting wires can increase the capacitance. Try to keep dropper lengths to a minimum and accessories, including block detection modules, as close to the track as possible, with short lengths of wire.

APPENDIX

NMRA Manufacturer Codes

This is an alphabetical list of the NMRA-issued ID codes for the manufacture of DCC decoders. It is updated regularly by the NMRA as and when new manufacturers apply for their code. Further details can be found on the NMRA DCC website.

Manufacturer (A–Z)	Binary	Hex	Decimal	Country
AE Electronic Ltd	10101001	0xA9	169	CHN
AMW	10011	0x13	19	AT
ANE Model Co. Ltd	101101	0x2D	45	TWN
Aristo-Craft Trains	100010	0x22	34	US
Arnold – Rivarossi	10101101	0xAD	173	DE
Atlas Model Railroad Products	1111111	0x7F	127	US
AuroTrains	10101010	0xAA	170	US/IT
Auvidel	1001100	0x4C	76	DE
AXJ Electronics	1101110	0x6E	110	CHN
Bachmann Trains	1100101	0x65	101	US
Benezan Electronics	1110010	0x72	114	ESP
Berros	1111001	0x7A	122	NL
BLOCKsignalling	10010100	0x94	148	UK
Bluecher-Electronic	111100	0x3C	60	DE
Blue Digital	11100001	0xE1	225	POL
Brelec	11111	0x1F	31	BE
BRAWA Modellspielwaren GmbH & Co.	10111010	0xBA	186	DE
Broadway Limited Imports, LLC	100110	0x26	38	US
Capecom	101111	0x2F	47	AU
CML Electronics Limited	1	0x01	1	UK
cmOS Engineering	10000010	0x82	130	AUS
Computer Dialysis France	1101001	0x69	105	FR
Con-Com GmbH	11001100	0xCC	204	AT
csikos-muhely	1111000	0x78	120	HUN
cT Elektronik	1110101	0x75	117	AT
CVP Products	10000111	0x87	135	US
Dapol Limited	10011010	0x9A	154	UK
DCCconcepts	100100	0x24	36	AU
DCC-Gaspar-Electronic	1111100	0x7C	124	HUN
DCC Supplies Ltd	110011	0x33	51	UK
DCC Train Automation	10010000	0x90	144	UK

Manufacturer (A–Z)	Binary	Hex	Decimal	Country
Desktop Station	10001100	0x8C	140	JP
Dietz Modellbahntechnik	1110011	0x73	115	DE
Digi-CZ	10011000	0x98	152	CZE
Digikeijs	101010	0x2A	42	NL
Digital Bahn	1000000	0x40	64	DE
Digitools Elektronika, Kft	1001011	0x4B	75	HUN
Digitrax	10000001	0x81	129	US
Digsight	11110	0X1E	30	CN
Doehler & Haas	1100001	0x61	97	DE
drM	10100100	0xA4	164	TWN
Educational Computer, Inc.	100111	0x27	39	US
Electronik & Model Produktion	100011	0x23	35	SE
Electronic Solutions Ulm GmbH	10010111	0x97	151	DE
Electroniscript, Inc.	1011110	0x5E	94	US
E-Modell	10000101	0x45	69	DE
Frateschi Model Trains	10000000	0x80	128	BRA
Fucik	10011110	0x9E	158	CZE
Gaugemaster	1000001	0x41	65	UK
Gebr. Fleischmann GmbH & Co.	10011011	0x9B	155	DE
Nucky	10011100	0x9C	156	JP
GFB Designs	101110	0x2E	46	UK
GooVerModels	1010001	0x51	81	BEL
Haber & Koenig Electronics GmbH (HKE)	1101111	0x6F	111	AT
HAG Modelleisenbahn AG	1010010	0x52	82	CHE
Harman DCC	1100010	0x62	98	UK
Hattons Model Railways	1001111	0x4F	79	UK
Heljan A/S	11100	0x1C	28	DK
Heller Modenlbahn	1000011	0x43	67	DE
Helvest Systems GmbH	10100111	0xA7	167	CH
HONS Model	1011000	0x58	88	HKG
Hornby Hobbies Ltd	110000	0x30	48	UK
Integrated Signal Systems	1100110	0x66	102	US
Intelligent Command Control	10000101	0x85	133	US
Joka Electronic	110001	0x31	49	DE
JMRI	10010	0x12	18	US
JSS-Elektronic	1010011	0x53	83	DE
KAM Industries	10110	0x16	22	US
KATO Precision Models	101000	0x28	40	JP
Kevtronics cc	1011101	0x5D	93	ZAF
Kreischer Datentechnik	10101	0x15	21	DE

Manufacturer (A–Z)	Binary	Hex	Decimal	Country
KRES GmbH	111010	0x3A	58	DE
Krois-Modell	110100	0x34	52	AT
Kuehn Ing.	10011101	0x9D	157	DE
LaisDCC	10000110	0x86	134	CHN
Lenz Elektronik GmbH	1100011	0x63	99	DE
LDH	111000	0x38	56	ARG
LGB (Ernst Paul Lehmann Patentwerk)	10011111	0x9F	159	DE
LSdigital	1110000	0x70	112	DE
LS Models Sprl	1001101	0x4D	77	BEL
Maison de DCC	10100110	0xA6	166	JP
Massoth Elektronik, GmbH	1111011	0x7B	123	DE
MAWE Elektronik	1000100	0x44	68	CH
MBTronik – PiN GITmBH	11010	0x1A	26	DE
MD Electronics	10100000	0xA0	160	DE
Mistral Train Models	11101	0x1D	29	BE
MoBaTron.de	11000	0x18	24	DE
Model Electronic Railway Group	10100101	0xA5	165	UK
Model Rectifier Corp.	10001111	0x8F	143	US
Model Train Technology	10101000	0xA8	168	US
Modelleisenbahn GmbH (formerly Roco)	10100001	0xA1	161	AT
Möllehem Gårdsproduktion	11111110	0x7E	126	SE
MTB Model	1001000	0x48	72	CZE
MTH Electric Trains, Inc.	11011	0x1B	27	US
MÜT GmbH	1110110	0x76	118	DE
MyLocoSound	1110100	0x74	116	AUS
N&Q Electronics	110010	0x32	50	ESP
NAC Services, Inc	100101	0x25	37	US
Nagasue System Design Office	1100111	0x67	103	JP
Nagoden	1101100	0x6C	108	JP
NCE Corporation	1011	0x0B	11	US
New York Byano Limited	1000111	0x47	71	HK
Ngineering	101011	0x2B	43	US
NMRA Reserved (for extended ID #s)	11101110	0xEE	238	US
Noarail	111111	0x3F	63	AUS
Nucky	10011100	0x9C	156	JP
NYRS	10001000	0x88	136	US
Opherline1	1101010	0x6A	106	FR
Passmann	101001	0x29	41	DE
Phoenix Sound Systems, Inc.	1101011	0x6B	107	US
PIKO	10100010	0xA2	162	DE

Manufacturer (A–Z)	Binary	Hex	Decimal	Country
PpP Digital	1001010	0x4A	74	ESP
Pojezdy.EU	1011001	0x59	89	CZE
Praecipuus	100001	0x21	33	CA
PRICOM Design	1100000	0x60	96	US
ProfiLok Modellbahntechnik GmbH	1111101	0x7D	125	DE
PSI – Dynatrol	1110	0x0E	14	US
Public Domain & Do-It-Yourself Decoders	1101	0x0D	13	-
QElectronics GmbH	110111	0x37	55	DE
QS Industries (QSI)	1110001	0x71	113	US
Railflyer Model Prototypes, Inc.	1010100	0x54	84	CAN
Railnet Solutions, LLC	1000010	0x42	66	US
Rails Europ Express	10010010	0x92	146	FR
Railstars Limited	1011011	0x5B	91	US
Ramfixx Technologies (Wangrow)	1111	0x0F	15	CA/US
Rampino Elektronik	111001	0x39	57	DE
Rautenhaus Digital Vertrieb	110101	0x35	53	DE
RealRail Effects	10001011	0x8B	139	US
Regal Way Co. Ltd	100000	0x20	32	HKG
Rock Junction Controls	10010101	0x95	149	US
Rocrail	1000110	0x46	70	DE
RR-Cirkits	1010111	0x57	87	US
S Helper Service	10111	0x17	23	US
Sanda Kan Industrial, Ltd	1011111	0x5F	95	HKG
Shourt Line	1011010	0x5A	90	US
SLOMO Railroad Models	10001110	0x8E	142	JP
Spectrum Engineering	1010000	0x50	80	US
SPROG-DCC	101100	0x2C	44	UK
T4T – Technology for Trains GmbH	10100	0x14	20	DE
Tam Valley Depot	111011	0x3B	59	US
Tams Elektronik GmbH	111110	0x3E	62	DE
Tawcrafts	1011100	0x5C	92	UK
TCH Technology	110110	0x36	54	US
Team Digital, LLC	11001	0x19	25	US
Tehnologistic (train-O-matic)	1001110	0x4E	78	ROM
The Electric Railroad Company	1001001	0x49	73	US
Throttle-Up (Soundtraxx)	10001101	0x8D	141	US
Train Control Systems	10011001	0x99	153	US
Train ID Systems	10001010	0x8A	138	US
TrainModules	111101	0x3D	61	HUN
Train Technology	10	0x02	2	BE

Manufacturer (A–Z)	Binary	Hex	Decimal	Country
TrainTech	1101000	0x68	104	NL
Trenes Digitales	1100100	0x64	100	ARG
Trix Modelleisenbahn	10000011	0x83	131	DE
Uhlenbrock GmbH	1010101	0x55	85	DE
Umelec Ing. Buero	10010011	0x93	147	CH
Viessmann Modellspielwaren GmbH	1101101	0x6D	109	DE
Wm. K. Walthers, Inc.	10010110	0x96	150	US
W. S. Ataras Engineering	1110111	0x77	119	US
Wangrow Electronics	1100	0x0C	12	US
Wekomm Engineering, GmbH	1010110	0x56	86	DE
WP Railshops	10100011	0xA3	163	CA
Zimo Elektronik	10010001	0x91	145	AT
ZTC	10000100	0x84	132	UK

Decoder Factory Reset Values

The standard reset is CV8 to 8. However, some manufacturers use different values and different CVs.

This list is by no means comprehensive, but it shows the difference between the manufacturers. It is a guide-line only; it is essential to read the decoder instructions before carrying out a reset.

Manufacturer	Manufacturer ID no. in CV8	CV	Value
Arnold/Rivarossi	173	CV8	8
Atlas Model Railroad Products	127	CV8	99
Bachmann Trains	101	CV8	8
Broadway Limited Imports (BLI)	38	CV8	8
Dapol	154	CV8	4
DCC Concepts	36	CV8	2 or 8
Digitrax	129	CV8	8
ESU (LokSound/LokPilot)	151	CV8	8
Gaugemaster	65	CV8	8
Hornby Hobbies Ltd	48	CV8	1
Kato Precision Industries	40	CV8	8
Lenz Elektronic	99	CV8	33
LGB	159	CV55	55
Model Rectifier Corp (MRC)	143	CV125	1
MTH	27	CV8	8
NCE Corporation	11	CV30	2
SoundTraxx: DSDLC, DSX, Tsunami	141	CV8	
SoundTraxx: DSDLC, DSX, Tsunami	141	CV30	2
Train Control Systems (TCS)	153	CV8	8

Manufacturer	Manufacturer ID no. in CV8	CV	Value
Train Control Systems (TCS)	153	CV30	2
Zimo Elektronic	145	CV8	8
ZTC	132	CV8	8

CV29

This chart is a guide to working out the required value of CV29 depending on the driving characteristics that need to be on. This is also applicable to other multi-use CVs; the only difference is the description of the characteristic for each bit.

CV29	Running Configurations							
Description	Direction 0=normal 1=reverse	Speed Step 0= 14 1= 28/128	Operation 0= DCC Only 1= DC and DCC	Railcom 0= off 1= on	Speed Curve 0= Preset 1= User Defined	Address 0= 2 Digit 1= 4 Digit	N/A	N/A
Bit	0	1	2	3	4	5	6	7
Value	1	2	4	8	16	32	64	128
On/Off								

Value of CV29 is the total of those bits which need to be switched on. Default setting is usually 6 (2-digit address and running on both DC and DCC).

Function Mapping

This chart is a guide to basic function mapping and should work with most makes of decoder. However, it is advisable to refer to the decoder manuals to check that the CVs relating to the function outputs use the same CVs as stated in this chart. NB: the red blocked out areas are unavailable for use.

Function	CV#	Default	Output (wire colour)													
			White	Yellow	Aux 1 (Green)	Aux 2 (Purple)	Aux 3 (Brown)	Aux 4 (Pink)	Aux 5	Aux 6	Aux 7	Aux 8	Aux 9	Aux 10	Aux 11	Aux 12
		Value	1	2	4	8	16	32	64	128	4	8	16	32	64	128
0F	33	1	X													
0R	34	2		X												
1	35	4			X											
2	36	8				X										
3	37	16					X									
4	38	32						X								
5	39	64							X							
6	40	128								X						
7	41	4									X					
8	42	8										X				
9	43	16											X			
10	44	32												X		
11	45	64													X	
12	46	128														X

Decoder Selection Chart

This chart has been put together with the latest information from manufacturers. It lists the current models available, with their model number, number of lighting functions (including logic where applicable), motor current, dimensions, and fitting style where the information is fully available.

The list is by no means complete as manufacturers add and remove decoders as production continues. The main intention with the chart is to show the fitting styles available and the dimensions, to make life easier when trying to find a suitable decoder for a locomotive.

The chart does not include function-only decoders or sound decoders.

Manufacturer	Model	Code	No. of function	No. of logic functions	Running amp	Max amp	L mm	W mm	H mm	Connector style
Dapol	Imperium 21 pin MTC	IMPERIUM1	6		1	2	18.00	16.00	3.00	21-pin direct
Dapol	Imperium Next18	IMPERIUM2	6		1	2	15.00	9.50	2.90	Next18 direct
Dapol	Imperium 21 pin MTC	IMPERIUM3	6	2	1	2	18.00	16.00	3.00	21-pin direct
Dapol	Imperium 6 pin Direct	IMPERIUM4	2		1	2	14.50	9.20	3.00	6-pin direct
Dapol	Imperium 6 pin Direct	IMPERIUM5	2		1	2	10.60	8.70	3.00	6-pin direct
Dapol	Imperium PluX22	IMPERIUM6	8		1	2	16.00	17.50	3.00	PluX22 direct
Dapol	Imperium 21-pin (Non-MTC)	IMPERIUM7	6		1	2	18.00	16.00	3.00	21-pin direct
DCC Concepts	Zen Black 21 pin MTC	DCD-Z218.6	6		0.75	1.1	22.00	16.00	5.00	21-pin direct + 8-pin-harness
DCC Concepts	Zen Black 21 pin MTC	DCD-ZN218.4.2	4	2	0.75	1.1	22.00	16.50	5.00	21-pin direct + 8-pin harness
DCC Concepts	Zen Black Universal	DCD-ZN360.6	6		0.75	1.1	16.00	14.00	5.00	8-pin direct
DCC Concepts	Zen Blue+ NEM651 6 pin Direct & Harness	DCD-ZN68.2	2		0.75	1.1	13.50	8.50	5.00	6-pin direct + 8-pin harness
DCC Concepts	Zen Blue+ NEM651 6 pin Direct	DCD-ZN6D.2	2		0.75	1.1	14.00	9.00	5.00	6-pin direct
DCC Concepts	Zen Blue+ 8 pin Nano	DCD-ZN8D.4	4		0.75	1.1	15.00	7.00	5.00	8-pin direct
DCC Concepts	Zen Black Super Thin Nano	DCD-ZN8H.2	2		0.75	1.1	16.00	9.00	3.80	8-pin harness
DCC Concepts	Zen Blue+ 8 pin Nano	DCD-ZN8H.nano	2		0.75	1.1	14.00	7.00	2.70	8-pin harness
DCC Concepts	Zen Black Midi	DCD-ZNM.HP6	6		1.5	3	30.00	18.30	6.00	8-pin harness
DCC Concepts	Zen Balck Mini	DCD-ZNmini.4	4		0.75	1.1	19.00	11.00	6.00	8-pin direct + 8-pin harness
DCC Concepts	Zen Blue+ Super Small Next18	DCD-ZNN18.4	4		0.75	1.1	15.00	11.00	3.80	Next18 direct
Digitrax	DG383AR	DG383AR	8		3	5	58.64	36.83		Aristocraft plug
Digitrax	DG583AR	DG583AR	8		5	10	56.64	36.83		Aristocraft plug
Digitrax	DG583S	DG583S	8		5	10	56.64	36.83		Screw terminal
Digitrax	DH126D	DH126D	2		1.5	2	27.28	17.08	6.60	Wires
Digitrax	DH126MT	DH126MT	2		1.5	2	27.28	17.08	6.60	21-pin direct

Manufacturer	Model	Code	No. of function	No. of logic functions	Running amp	Max amp	L mm	W mm	H mm	Connector style
Digitrax	DH126P	DH126P	2		1.5	2	27.28	17.08	6.60	8-pin harness long
Digitrax	DH126PS	DH126PS	2		1.5	2	27.28	17.08	6.60	8-pin harness short
Digitrax	DH165A0	DH165A0	6		1.25	2	73.07	16.93	4.40	Board replacement
Digitrax	DH165IP	DH165IP	6		1.5	2	17.02	26.67	6.35	8-pin direct
Digitrax	DH165K0	DH165K0	6		1.25	2	73.07	16.93	4.40	Board replacement
Digitrax	DH165K1A	DH165K1A	6		1.25	2	73.07	16.93	4.40	Board replacement
Digitrax	DH165L0	DH165L0	6		1.5	2	26.67	17.02	6.35	Board replacement
Digitrax	DH165Q1	DH165Q1	6		1.5	2	73.07	16.93	4.40	Board replacement
Digitrax	DH166D	DH166D	6		1.5	2	27.28	17.08	6.60	Wires
Digitrax	DH166MT	DH166MT	6		1.5	2	27.28	17.08	6.60	21-pin direct
Digitrax	DH166P	DH166P	6		1.5	2	27.28	17.08	6.60	8-pin harness long
Digitrax	DH166PS	DH166PS	6		1.5	2	27.28	17.08	6.60	8-pin harness short
Digitrax	DH465	DH465	6		4	6	51.80	15.24	2.79	Wires
Digitrax	DN126M2	DN126M2	6		1	1.25	15.79	10.63	2.72	Board replacement
Digitrax	DN123K3	DN123K3	2		1.25	2	34.04	9.91	2.54	Board replacement
Digitrax	DN136D	DN136D	3		1	1.5	13.90	10.30	5.00	Wires
Digitrax	DN136PS	DN136PS	3		1	1.5	13.90	10.30	5.00	8-pin harness short
Digitrax	DN143K2	DN143K2	4		1	1.5				4-part decoder incl. light boards
Digitrax	DN146IP	DN146IP	4		1	1.5	29.43	9.81	2.98	8-pin direct
Digitrax	DN163A0	DN163A0	6		1	1.25	76.24	9.50	3.99	Board replacement
Digitrax	DN163A1	DN163A1	6		1	1.25	98.43	9.50	3.99	Board replacement
Digitrax	DN163A3	DN163A3	6		1	1.25	98.43	9.50	3.99	Board replacement
Digitrax	DN163A4	DN163A4	6		1.5	2	98.30	9.50	4.00	Board replacement
Digitrax	DN163K0A	DN163K0A	6		1	1.25	56.30	13.93	2.58	Board replacement
Digitrax	DN163K0B	DN163K0B	6		1	1.25	56.30	13.93	2.58	Board replacement
Digitrax	DN163K0D	DN163K0D	6		1	1.25	56.30	13.93	2.58	Board replacement

Manufacturer	Model	Code	No. of function	No. of logic functions	Running amp	Max amp	L mm	W mm	H mm	Connector style
Digitrax	DN163K1C	DN163K1C	6		1	1.25	78.99	10.82	4.95	Board replacement
Digitrax	DN163K1D	DN163K1D	6		1	1.25	78.99	10.82	4.95	Board replacement
Digitrax	DN163K2	DN163K2	6		1	1.25	80.77	8.99	3.25	Board replacement
Digitrax	DN163K4A	DN163K4A	6		1	1.25	59.72	10.44	1.55	Board replacement
Digitrax	DN163K4B	DN163K4B	6		1	1.25	59.72	10.44	1.55	Board replacement
Digitrax	DN163L0A	DN163L0A	6		1	1.25	53.85	8.89	1.12	Board replacement
Digitrax	DN163M0	DN163M0	6		1	1.5	54.81	14.73	3.43	Board replacement
Digitrax	DN166I0	DN166I0	6		1.5	2	88.90	9.32	3.42	Board replacement
Digitrax	DN166I1A	DN166I1A	6		1.5	2	54.99	11.98	2.40	Board replacement
Digitrax	DN166I1B	DN166I1B	6		1.5	2	54.99	11.98	2.40	Board replacement
Digitrax	DN166I1C	DN166I1C	6		1.5	2	54.99	11.98	2.40	Board replacement
Digitrax	DN166I1D	DN166I1D	6		1.5	2	54.99	11.98	2.40	Board replacement
Digitrax	DN166I2	DN166I2	6		1.5	2	54.99	11.98	2.40	Board replacement
Digitrax	DN166I2B	DN166I2B	6		1.5	2	54.99	11.98	2.40	Replacement board
Digitrax	DN166PS	DN166PS	6		1	1.5	22.16	10.28	5.10	8-pin harness short
Digitrax	DZ123	DZ123	2		1	2	13.97	9.14	3.30	Wires
Digitrax	DZ123M0	DZ123M0	2		1	2	33.50	8.13	3.00	Board replacement
Digitrax	DZ123MK0	DZ123MK0	2		1	1.25	59.87	10.69	2.21	Board replacement
Digitrax	DZ123MK1	DZ123MK1	2		1	1.25	69.14	10.54	2.74	Board replacement
Digitrax	DZ123PS	DZ123PS	2		1	2	14.50	9.80	3.20	8-pin harness short
Digitrax	DZ123Z0	DZ123Z0	2		1	1.25	32.92	6.88	2.51	Board replacement
Digitrax	DZ126	DZ126	2		1	1.5	11.56	9.37	3.20	Wires
Digitrax	DZ126IN	DZ126IN	2		1	1.5	10.61	9.05	2.67	6-pin direct
Digitrax	DZ126PS	DZ126PS	2		1	1.5	11.56	9.37	3.20	8-pin harness short

Manufacturer	Model	Code	No. of function	No. of logic functions	Running amp	Max amp	L mm	W mm	H mm	Connector style
Digitrax	DZ126T	DZ126T	2		1	1.25	14.00	7.13	3.25	Wires
Digitrax	DZ126Z1	DZ126Z1	2		1	1.5	11.56	9.37	3.20	Board replacement
Digitrax	DZ146	DZ146	4		1	1.5	14.00	10.23	3.77	Wires
Digitrax	DZ146IN	DZ146IN	4		1	1.5	14.20	9.80	3.48	6-pin direct
Digitrax	DZ146PS	DZ146PS	4		1	1.5	14.00	10.23	3.77	8-pin harness short
ESU	LokPilot 5 Basic 8-pin NEM652	59020	4	8	0.9	1	21.40	15.50	4.50	8-pin harness
ESU	LokPilot 5 Basic 21MTC	59029	4	8	0.9	1	21.40	15.50	4.50	21-pin direct
ESU	LokPilot 5 DCC/MM/SX/M4 8 pin	59610	4		1.5	2	21.40	15.50	4.50	8-pin harness
ESU	LokPilot 5 DCC/MM/SX/M4 PluX22	59612	10	2	1.5	2	21.40	15.50	4.50	PluX22 direct
ESU	LokPilot 5 DCC/MM/SX/M4 6-pin	59616	4		1.5	2	21.40	15.50	4.50	6-pin harness
ESU	LokPilot 5 DCC/MM/SX/M4 21MTC	59619	10	2	1.5	2	21.40	15.50	4.50	21-pin direct
ESU	LokPilot 5 DCC/MM/SX/M4 21MTC MKL	59649	10	2	1.5	2	21.40	15.50	4.50	21-pin direct
ESU	LokPilot 5 DCC 8-pin NEM652	59620	4		1.5	2	21.40	15.50	4.50	8-pin harness
ESU	LokPilot 5 DCC PluX22 NEM658	59622	10		1.5	2	21.40	15.50	4.50	PluX22 direct
ESU	LokPilot 5 DCC 6-pin NEM651	59626	4		1.5	2	21.40	15.50	4.50	6-pin direct
ESU	LokPilot 5 DCC 21MTC NEM660	59629	10		1.5	2	21.40	15.50	4.50	21-pin direct
ESU	LokPilot 5 DCC 21MTC MKL	59659	10		1.5	2	21.40	15.50	4.50	21-pin direct
ESU	LokPilot 5 micro DCC/MM/SX 8-pin	59810	4		0.75	0.75	8.00	7.00	2.40	8-pin direct
ESU	LokPilot 5 micro DCC/MM/SX PluX16	59814	6		0.75	0.75	13.00	9.20	2.40	PluX16 direct
ESU	LokPilot 5 micro DCC/MM/SX 6-pin	59816	4		0.75	0.75	8.00	7.00	2.40	6-pin harness
ESU	LokPilot 5 micro DCC/MM/SX 6-pin Direct	59817	4		0.75	0.75	8.00	7.00	2.40	6-pin direct
ESU	LokPilot 5 micro DCC/MM/SX/M4 Next18	59818	6		0.75	0.75	13.00	9.00	2.40	Next18 direct
ESU	LokPilot 5 micro DCC/MM/SX 6-pin Direct Right Angle	59837	4		0.75	0.75	8.00	7.00	2.40	6-pin direct (right angled)
ESU	LokPilot 5 micro DCC 8-pin NEM652	59820	4		0.75	0.75	8.00	7.00	2.40	8-pin harness

Manufacturer	Model	Code	No. of function	No. of logic functions	Running amp	Max amp	L mm	W mm	H mm	Connector style
ESU	LokPilot 5 micro DCC PluX16	59824	6		0.75	0.75	13.00	9.20	2.40	PluX16 direct
ESU	LokPilot 5 micro DCC 6-pin NEM651	59826	4		0.75	0.75	8.00	7.00	2.40	6-pin harness
ESU	LokPilot 5 micro DCC 6-pin NEM651 Direct	59827	4		0.75	0.75	8.00	7.00	2.40	6-pin direct
ESU	LokPilot 5 micro DCC Next18	59828	6		0.75	0.75	13.00	9.20	2.40	Next18 direct
ESU	LokPilot 5 micro DCC 6-pin Direct Right Angle	59857	4		0.75	0.75	8.00	7.00	2.40	6-pin direct (right angled)
ESU	LokPilot micro Direct for Narrow Hood Units	54640	6		0.75	0.75	66.00	8.20	4.50	Board replacement
ESU	LokPilot 5 L DCC/MM/SX/ M4 Terminal Block	59315	11	6	3	3	50.00	25.00	12.00	Terminal block
ESU	LokPilot 5 L DCC/MM/SX/ M4 with Pins	59325	11	6	3	3	50.00	25.00	12.00	Pins
Fleischmann	Next18 NEM662 Direct Plug	FM685101	4		0.7	0.7	15.00	9.50	2.80	Next18 direct
Fleischmann	6 pin NEM651 Direct Plug	FM685305	4		0.7	0.7	13.00	9.00	2.60	6-pin direct
Fleischmann	6 pin NEM651 Wired 80mm Harness	FM685404	4		0.8	0.8	14.00	9.00	2.50	6-pin harness long
Fleischmann	6 pin NEM651 Wired 35mm Harness	FM685504	4		0.8	0.8	14.00	9.00	2.50	6-pin harness short
Fleischmann	6 pin NEM651 Direct Plug	FM686101	4		0.8	0.8	14.00	9.00	2.50	6-pin direct
Fleischmann	8 pin NEM652 Wired Harness	FM686201	4		0.8	0.8	14.00	9.00	2.50	8-pin harness
Fleischmann	6 pin NEM651 Wired Harness	FM687403	4		1	1	20.00	11.00	3.50	6-pin harness
Gaugemaster	OMNI 4fn Next18	DCC18	4		1	1.5	15.00	9.00	3.00	Next18 direct
Gaugemaster	Kato Class 800 Set	DCC89			1	1.5				Class 800 decoder set
Gaugemaster	Ruby Series 2fn Standard 8 pin	DCC90	2		1.5	2	27.00	17.00	7.00	8-pin harness
Gaugemaster	Ruby Series 2fn Standard 21 pin	DCC91	2		1.5	2	21.00	16.00	4.00	21-pin direct
Gaugemaster	Ruby Series 2fn Small 8 pin	DCC92	2		1	1.5	12.00	10.00	3.00	8-pin harness
Gaugemaster	Ruby Series 2fn Small 6 pin	DCC93	2		1	1.5	10.00	11.00	3.00	6-pin direct
Gaugemaster	Ruby Series 6fn Pro 8 pin	DCC94	6		1.5	2	27.00	17.00	7.00	8-pin harness short
Gaugemaster	Ruby Series 6fn Pro 21 pin	DCC95	6		1.5	2	21.00	16.00	4.00	21-pin direct

Manufacturer	Model	Code	No. of function	No. of logic functions	Running amp	Max amp	L mm	W mm	H mm	Connector style
Lenz	Standard+ v2	10231-02	4		1	1	25.00	15.00	3.80	8-pin harness
Lenz	Silvermini+ with wires	10310-02	2		0.5	0.5	10.60	7.50	2.60	Wires
Lenz	Silvermini+	10311-02	2		0.5	0.5	13.00	7.50	2.80	6-pin direct
Lenz	Silvermini+ V2 with wires	10310-03	4		0.5	0.8	10.60	7.50	2.60	Wires
Lenz	Silvermini+ V2	10311-03	4		0.5	0.8	10.60	7.50	2.60	6-pin direct
Lenz	Silver+ PluX12	10312-01	4		0.75	1	20.00	11.00	4.00	PluX12 direct
Lenz	Silver + PluX22	10322-01	9		0.75	1	22.00	15.00	6.00	PluX22 direct
Lenz	Silver+ Next18	10318-01	8		0.6	1	15.00	9.50	2.90	Next18 direct
Lenz	Silver+ 21	10321-01	4		1	1.8	20.60	15.70	4.00	21-pin direct
Lenz	Silver+ Direct	10330-01	4		1	1.8	19.20	13.00	3.60	8-pin direct
Lenz	Gold+	10433-01	5		1	1.8	22.80	16.70	4.90	Wires
NCE	N14A3 for Atlas & Intermountain	5240181	4		0.75	0.75	66.00	8.90	2.00	Board replacement
NCE	Next18 Direct Plug	5240178	4		0.75	1.2	15.00	10.00	2.00	Next18 direct
NCE	N12 6 pin	5240160	4		0.75	0.75	19.41	8.77	2.65	6-pin direct
NCE	N14K2 for Kato, Lifelike & Walthers	5240169	4		0.75	0.75				Board replacement
NCE	N14K1 for Kato & Athearn	5240167	4		0.75	0.75				Board replacement
NCE	D13WP 8 pin wired	5240177	4		0.75	1.2	26.00	16.50	4.70	8-pin harness
NCE	D13J 9 Pin Direct Plug	5240174	4		0.75	1.2	26.00	16.50	4.70	9-pin JST Socket with 8-pin harness
NCE	D13W wired	5240171	4		0.75	1.2	26.00	16.50	4.70	Wires
NCE	D16MTC 21 pin	5240156	8		0.75	1.2	23.00	16.00	4.70	21-pin direct
NCE	NMP15 for Atlas MP15	5240138	4		1	1.25				Board replacement
NCE	H15/16-44 for Atlas H15-44 & H16-44	5240159	4		0.7	1				Board replacement
NCE	D13NHP 8 pin Decoder with Keep Alive	5240147	4		1.3	2				8-pin harnes + Keep Alive
NCE	D13NHJ 9 pin Decoder with Keep Alive	5240146	4		1.3	2				9-pin JST socket with 8-pin harness + Keep Alive
NCE	N12K0B Drop In Decoder for Kato F3 A & B Units	5240143	2		1	1.25				Board replacement
NCE	N12A2 Drop In Decoder for Atlas GP7/9/30/35	5240142	2		1	1.25				Board replacement
NCE	Bach-Dsl HO Scale Decoder for Bachmann	5240139	2		1	1.25				Board replacement
NCE	NAVO Drop In Decoder for Atlas VO-1000	5240137	2		1	1.25				Board replacement
NCE	N14SRP 8 pin wired Decoder	5240132	4		1	1.25	18.00	8.60	3.20	8-pin harness

Manufacturer	Model	Code	No. of function	No. of logic functions	Running amp	Max amp	L mm	W mm	H mm	Connector style
NCE	N14SR wired Decoder	5240131	4		1	1.25	29.21	10.16	3.05	Wires
NCE	N14IP 8 pin Decoder	5240128	4		1	1.25	29.21	10.16	2.79	8-pin direct
NCE	N12A0e Drop In Decoder Intermountain Tunnel	5240127	4		1.3	2				Board replacement
NCE	N12A1 Drop In Decoder for Atlas	5240126	4		1	2				Board replacement
NCE	N12A0 Drop In Decoder for Atlas	5240120	4		1	1.25	67.31	9.40	3.05	Board replacement
NCE	N12SR N Scale wired Decoder	5240119	4		1	1	18.00	8.60	3.20	Wires
NCE	D808 Decoder	5240112	8		8	30	96.00	37.00	15.00	Screw terminal
NCE	D408 Decoder	5240111	6		4	10	59.00	31.00	9.50	Screw terminal
NCE	SW9-SR Decoder for Walthers/Life Like	5240110	3		1	1.25			2.54	Board replacement
NCE	DA-SR Solder In Decoder	5240106	5		1.3	2	72.00	17.00	3.20	Board replacement
NCE	D14SR 6 pin Decoder	5240103	6		1.3	2	21.00	16.00		8-pin direct
Roco	6 pin decoder with angled pins	10887	2		0.7	0.7	13.00	9.00	2.60	6-pin direct (right angled)
Roco	8 pin NEM652 Wired Harness	10894	4		1	1	20.00	11.00	4.00	8-pin harness
Roco	Plux16 Direct Plug Decoder	10895			1	1	20.00	11.00	4.00	PluX16 direct
Roco	PluX22 Direct Plug Decoder	10896			1.2	1.2	22.00	15.00	4.00	PluX22 direct
SoundTraxx	HO Scale 21 Pin Decoder	MC1H104P21	4		1	1	25.10	15.60	4.70	21-pin direct
SoundTraxx	6 pin Mobile Decoder	MC1Z102P6	2		1	1	13.00	9.00	3.00	6-pin direct
SoundTraxx	Wired Mobile Decoder	MC1Z102SQ	2		1	1	13.00	9.00	3.00	Wires
SoundTraxx	8 pin Mobile Decoder	MC1H102P8	2		1	1	17.00	17.00	7.00	8-pin direct
SoundTraxx	Drop In Decoder for Atlas & Intermountain	MC2H104AT	4		1	1	74.00	17.00	3.00	Board replacement
SoundTraxx	8 Pin Wired Mobile Decoder	MC2H104OP	4		1	1	25.00	16.00	6.00	8-pin harness
SoundTraxx	Wired Mobile Decoder	MC2H104P9	4		1	1	25.00	16.00	6.00	Wires
TCS	A4X Atlas, Athearn, Kato, Bachmann	1000	6		1.3	2	72.64	17.27	3.68	Board replacement
TCS	A6X Atlas, Athearn, Kato, Bachmann	1001	4		1.3	2	72.64	17.27	3.68	Board replacement
TCS	ALD4	1302	4		1	2	77.09	8.89	3.30	Board replacement
TCS	AMD4	1029	4		1	2	71.88	####	3.30	Board replacement
TCS	AS6	1416	6		1	2	####	22.64	2.71	Board replacement

Manufacturer	Model	Code	No. of function	No. of logic functions	Running amp	Max amp	L mm	W mm	H mm	Connector style
TCS	ASD4	1279	4		1	2	58.65	9.02	2.26	Board replacement
TCS	AZL1D4	1560	4		1.3	2	42.00	6.90	2.54	Board replacement
TCS	AZL4	1550	4		1.3	2	33.00	6.60	2.54	Board replacement
TCS	CN-GP	1286	4		1	2	0.00	0.00	0.00	Wires
TCS	CN-Series	1278	2		1	2	19.90	9.17	2.60	Wires
TCS	DP2-LL	1412	2		1	2	17.83	2.04	4.06	8-pin direct
TCS	DP2X	1028	2		1	2	12.04	17.83	3.05	8-pin direct
TCS	DP2X-UK	1287	2		1	2	12.04	17.83	3.05	8-pin direct
TCS	DP5	1335	5		2	2	12.92	16.15	2.64	8-pin direct
TCS	EU821 21 pin Decoder	1674	8		1	2	20.24	15.58	5.00	21-pin direct
TCS	EUN651 6 pin Direct Plug	1298	2		1	2	13.82	8.76	2.84	6-pin direct
TCS	EUN651P-18	1373	2		1	2	13.82	8.76	2.84	6-pin harness short
TCS	EUN651P-30	1937	2		1	2	13.82	8.76	2.84	6-pin harness long
TCS	G8	1303	8		4	8	76.52	37.07	13.36	Terminal block
TCS	IMD4	1327	4		1	2	71.88	8.92	3.30	Board replacement
TCS	IMD4-W	1672	4		1	2	61.55	9.02	2.26	Board replacement
TCS	IMF4	1328	4		1	2	11.93	54.81	4.54	Board replacement
TCS	IMF4-NF	1551	4		1	2	11.93	54.81	4.54	Board replacement
TCS	IMFP4	1329	4		1	2	11.93	54.81	4.54	Board replacement
TCS	IMFP4-NF	1552	4		1	2	11.93	54.81	4.54	Board replacement
TCS	IMFTA4	1330	4		1	2	11.93	54.81	4.54	Board replacement
TCS	IMFTB4	1331	4		1	2	11.93	54.81	4.54	Board replacement
TCS	K0D8-A	1332	8		1	2	64.26	13.77	3.94	Board replacement
TCS	K0D8-B	1333	8		1	2	64.26	13.77	3.94	Board replacement
TCS	K0D8-C	1338	8		1	2	64.26	13.77	3.94	Board replacement
TCS	K0D8-D	1339	8		1	2	64.26	13.77	3.94	Board replacement
TCS	K0D8-E	1481	8		1	2	64.26	13.76	3.94	Board replacement

Manufacturer	Model	Code	No. of function	No. of logic functions	Running amp	Max amp	L mm	W mm	H mm	Connector style
TCS	K0D8-F Kato N-Scale EMD	2015	6		1	2	59.53	12.46	2.46	Board replacement
TCS	K0D8-G	2016	8		1	2	54.32	13.67	1.73	Board replacement
TCS	K1D4	1293	4		1	2	68.70	9.64	1.50	Board replacement
TCS	K1D4-NC	1318	4		1	2	68.70	9.64	1.50	Board replacement
TCS	K2D4	1294	4		1	2	67.00	8.95	1.73	Board replacement
TCS	K3D3	1295	3		1	2	32.02	10.97	1.73	Board replacement
TCS	K4D6 Kato N-Scale EMD	1414	6		1	2	59.53	12.46	2.46	Board replacement
TCS	K5D7 Drop in For F40PH	1489	7		1	2	68.80	13.83	1.81	Board replacement
TCS	K6D4 Kato FEF	1556	4		1	2	59.77	10.49	1.81	Board replacement
TCS	KAM4	1485	4		1	2	24.00	16.70	8.38	Wires
TCS	KAM4-LED	1479	4		1	2	24.00	16.70	8.38	Wires
TCS	KAM4P-MH	1487	4		1	2	24.00	16.70	8.38	8-pin harness
TCS	KAT24 with Stay Alive	1465	4		1	2	33.40	16.50	8.90	Wires
TCS	KAT26 with Stay Alive	1466	6		1	2	33.40	16.50	8.90	Wires
TCS	L1D4	1413	4		1	2	52.30	8.87	3.35	Board replacement
TCS	LL8	1343	8		1	2	50.51	17.64	3.40	Board replacement
TCS	LL8-LED	1402	8		1	2	50.51	17.64	3.40	Board replacement
TCS	M1 Miniture 2 Function	1006	2		1	2	14.40	9.12	3.43	Wires
TCS	M1P-3.5″	1386	2		1	2	14.40	9.12	3.43	8-pin harness 3.5′
TCS	M1P-3.5″R	1387	2		1	2	14.40	9.12	3.43	8-pin harness 3.5′R
TCS	M1P-3.5″UK	1390	2		1	2	14.40	9.12	3.43	8-pin harness 3.5′UK
TCS	M1P-5″	1388	2		1	2	14.40	9.12	3.43	8-pin harness 5′
TCS	M4 Miniture 4 Function	1011	4		1	2	14.40	9.12	3.43	Wires
TCS	M4P-3.5″UK	1396	4		1	2	14.40	9.12	3.43	8-pin harness 3.5′UK
TCS	M4P-3.5″	1392	4		1	2	14.40	9.12	3.43	8-pin harness 3.5′
TCS	M4P-3.5″R	1393	4		1	2	14.40	9.12	3.43	8-pin harness 3.5′R
TCS	M4P-5″	1394	4		1	2	14.40	9.12	3.43	8-pin harness 5′

Manufacturer	Model	Code	No. of function	No. of logic functions	Running amp	Max amp	L mm	W mm	H mm	Connector style
TCS	MC2 2 Function Mini	1013	2		1	2	18.50	10.60	4.80	Wires
TCS	MC4	1017	4		1	2	18.02	10.63	4.85	Wires
TCS	MP15N	1030	4		1	2	52.95	9.01	2.41	Board replacement
TCS	MT1500	1549	2		1.3	2	12.95	9.90	2.54	Board replacement
TCS	T1-LED	1484	2		1.3	2	24.61	16.79	4.90	Wires
TCS	T4-LED	1482	4		1.3	2	24.61	16.79	4.90	Wires
TCS	T4X	1024	4		1.3	2	24.61	16.79	4.90	Wires
TCS	VO-1000	1031	4		1	2	76.96	9.02	3.43	Board replacement
TCS	Z2 2 Function Micro Mini	1296	2		1	1	12.95	6.90	2.79	Wires
TCS	Z2P-1"	1378	2		1	2	12.95	6.90	2.79	8-pin harness 1'
TCS	Z2P-3.5"	1379	2		1	2	12.95	6.90	2.79	8-pin harness 3.5'
TCS	Z2P-3.5" UK	1380	2		1	2	12.95	6.90	2.79	8-pin harness 3.5'
trainOmatic	Lokommander II Mini W6P	2010207	4		1	1	19.50	11.00	3.30	6-pin harness
trainOmatic	Lokommander II Mini M21	2010208	6		1	1	20.00	15.30	5.00	21-pin direct
trainOmatic	Lokommander II Mini M21P	2010228	6		1	1	20.00	15.30	5.00	21-pin direct
trainOmatic	Lokommander II Mini M21S	2010209	6		1	1	20.00	15.30	5.00	21-pin direct
trainOmatic	Lokommander II Mini P12	2010210	6		1	1	19.50	11.00	3.00	PluX12 direct
trainOmatic	Lokommander II Mini P16	2010211	6		1	1	19.50	11.00	3.00	PluX16 direct
trainOmatic	Lokommander II Mini W8P	2010212	4		1	1	19.50	11.00	3.30	8-pin harness
trainOmatic	Lokommander II Micro N18	2010216	6		1	1	14.20	9.20	3.00	Next18 direct
trainOmatic	Lokommander II Mini P22	2010217	10		1	1	20.50	15.00	3.20	PluX22 direct
trainOmatic	Lokommander II Mini W22	2010218	10		1	1	20.50	15.00	3.20	8-pin harness
trainOmatic	Lokommander II Mini W22M	2010229	10		1	1	20.50	15.00	3.20	Wires
trainOmatic	Lokommander II Micro 6P	2010220	4		1	1	14.00	9.00	3.30	6-pin direct
trainOmatic	Lokommander II Micro 6P90	2010221	4		1	1	14.00	9.00	3.30	6-pin direct (right angled)
trainOmatic	Lokommander II Micro 6P90R	2010227	4		1	1	14.00	9.00	3.30	6-pin direct (right angled)

Manufacturer	Model	Code	No. of function	No. of logic functions	Running amp	Max amp	L mm	W mm	H mm	Connector style
trainOmatic	Lokommander II Micro W6P	2010222	4		1	1	14.00	9.00	3.30	6-pin harness
trainOmatic	Lokommander II Micro W	2010223	4		1	1	14.00	9.00	3.30	Wires
Uhlenbrock	Intellidrive 2 Basic	74320	1		0.65	1	19.00	14.00	3.50	8-pin harness
Uhlenbrock	Intellidrive 2 Deluxe 21MTC	75335	6		1.2	2	20.50	15.40	4.60	21-pin direct
Uhlenbrock	Intellidrive 2 8 pin	74120	2		1.2	2	20.00	11.00	4.60	8-pin harness
Uhlenbrock	Intellidrive 2 Deluxe 8 pin	74125	2		1.2	2	20.00	11.00	4.60	8-pin harness
Uhlenbrock	Intellidrive 2 PluX16	74150	2		1.2	2	20.00	11.00	3.80	PluX16 direct
Uhlenbrock	Intellidrive 2 Deluxe PluX16	74155	2		1.2	2	20.00	11.00	3.80	PluX16 direct
Uhlenbrock	Intellidrive 2 PluX22	74560	7		1.2	2	22.00	15.00	3.80	PluX22 direct
Uhlenbrock	Intellidrive 2 Deluxe PluX22	74570	7		1.2	2	22.00	15.00	3.80	PluX22 direct
Viessmann	HO Scale Wired Decoder	5244	4		1	1.8	25.00	15.40	3.30	Wires
Viessmann	HO Scale 8 pin Decoder	5245	4		1	1.8	25.00	15.40	3.30	8-pin harness
Viessmann	DCC/Selextrix DH10A Decoder	52521	4		1	1	14.30	9.20	8.00	6-pin ribbon cable harness
Viessmann	Decoder with Cable	5296	2		0.5	0.8	11.50	9.50	2.10	Wires
Viessmann	6 pin decoder	5297	2		0.5	0.8	15.60	9.50	2.10	6-pin direct
Viessmann	Next18 Decoder	5298			0.5	0.8	15.50	9.50	2.10	Next18 direct
Zimo	MX600	MX600	4		0.8	1.5	25.00	11.00	2.00	Wires
Zimo	MX600R	MX600R	4		0.8	1.5	25.00	11.00	2.00	8-pin harness
Zimo	MX600P12	MX600P12	4		0.8	1.5	25.00	11.00	2.00	PluX12 direct
Zimo	MX615	MX615	4		0.5	1	8.20	5.70	2.00	Wires
Zimo	MX615R	MX615R	4		0.5	1	8.20	5.70	2.00	8-pin harness
Zimo	MX615F	MX615F	4		0.5	1	8.20	5.70	2.00	6-pin harness
Zimo	MX615N	MX615N	4		0.5	1	8.20	5.70	2.00	6-pin direc
Zimo	MX616	MX616	6		0.7	1.5	8.00	8.00	2.40	Wires
Zimo	MX616F	MX616F	6		0.7	1.5	8.00	8.00	2.40	6-pin harnes
Zimo	MX616N	MX616N	6		0.7	1.5	8.00	8.00	2.40	6-pin direct
Zimo	MX616R	MX616R	6		0.7	1.5	8.00	8.00	2.40	8-pin harness
Zimo	MX617	MX617	6		0.8	1.5	13.00	9.00	2.50	Wires
Zimo	MX617F	MX617F	6		0.8	1.5	13.00	9.00	2.50	6-pin harness
Zimo	MX617N	MX617N	6		0.8	1.5	13.00	9.00	2.50	6-pin direct
Zimo	MX617R	MX617R	6		0.8	1.5	13.00	9.00	2.50	8-pin harness
Zimo	MX618	MX618N18	4	2	0.8	1.5	15.00	9.50	2.80	Next18 direct
Zimo	MX622	MX622	4		0.8	1.5	14.00	9.00	2.50	Wires
Zimo	MX622F	MX622F	4		0.8	1.5	14.00	9.00	2.50	6-pin harness

Manufacturer	Model	Code	No. of function	No. of logic functions	Running amp	Max amp	L mm	W mm	H mm	Connector style
Zimo	MX622N	MX622N	4		0.8	1.5	14.00	9.00	2.50	6-pin direct
Zimo	MX622R	MX622R	4		0.8	1.5	14.00	9.00	2.50	8-pin harness
Zimo	MX623	MX623	4	2	0.8	1.5	20.00	8.50	3.50	Wires
Zimo	MX623F	MX623F	4	2	0.8	1.5	20.00	8.50	3.50	6-pin harness
Zimo	MX623P12	MX623P12	4	2	0.8	1.5	20.00	8.50	3.50	PluX12 direct
Zimo	MX623R	MX623R	4	2	0.8	1.5	20.00	8.50	3.50	8-pin harness
Zimo	MX630	MX630	6	2	1	2.5	20.00	11.00	3.50	Wires
Zimo	MX630F	MX630F	6	2	1	2.5	20.00	11.00	3.50	6-pin harness
Zimo	MX630P16	MX630P16	6	2	1	2.5	20.00	11.00	3.50	PluX12 direct
Zimo	MX630R	MX630R	6	2	1	2.5	20.00	11.00	3.50	8-pin harness
Zimo	MX633	MX633	10	2	1.2	2.5	22.00	15.00	3.50	Wires
Zimo	MX633P16	MX633P16	10	2	1.2	2.5	22.00	15.00	3.50	PluX16 direct
Zimo	MX633P22	MX633P22	10	2	1.2	2.5	22.00	15.00	3.50	PluX22 direct
Zimo	MX633R	MX633R	10	2	1.2	2.5	22.00	15.00	3.50	8-pin harness
Zimo	MX633F	MX633F	10	2	1.2	2.5	22.00	15.00	3.50	6-pin harness
Zimo	MX634	MX634	6	2	1.2	2.5	20.50	15.50	3.50	Wires
Zimo	MX634D	MX634D	6	2	1.2	2.5	20.50	15.50	3.50	21-pin direct
Zimo	MX634R	MX634R	6	2	1.2	2.5	20.50	15.50	3.50	8-pin harness
Zimo	MX635	MX635	10	2	1.8	2.5	25.00	15.00	3.50	Wires
Zimo	MX635R	MX635R	10	2	1.8	2.5	25.00	15.00	3.50	8-pin harness
Zimo	MX636D	MX636D	8	2	1.8	2.5	26.00	15.00	3.50	21-pin direct
Zimo	MX637P22	MX637P22	10	2	1.2	2.5	22.00	15.00	3.50	PluX22 direct
Zimo	MX638D	MX638D	6	2	1.2	2.5	20.50	15.50	3.50	21-pin direct
Zimo	M330	M330	10	2	1.2	1.2	30	15.3	2.2	Wires
Zimo	M330R	M330R	10	2	1.2	1.2	30	15.3	2.2	8-pin harness
Zimo	MN330P22	MN330P22	10	2	1.2	1.2	30	15.3	2.2	PluX22 direct
Zimo	MN340C	MN340C	4	6	1.2	1.2	28.6	15.3	2.5	21-pin direct
Zimo	MN340D	MN340D	8	2	1.2	1.2	28.6	15.3	2.5	21-pin direct
Zimo	MN300	MN300	6	2	1	1	17.6	10.5	3.1	Wires
Zimo	MN300R	MN300R	6	2	1	1	17.6	10.5	3.1	8-pin harness
Zimo	MN300F	MN300F	6	2	1	1	17.6	10.5	3.1	6-pin harness
Zimo	MN300P16	MN300P16	6	2	1	1	17.6	10.5	3.1	PluX16 direct
Zimo	MN170	MN170	6	2	0.7	0.7	12	8.6	2.3	Wires
Zimo	MN170R	MN170R	6	2	0.7	0.7	12	8.6	2.3	8-pin harness
Zimo	MN170F	MN170F	6	2	0.7	0.7	12	8.6	2.3	6-pin harness
Zimo	MN170N	MN170N	6	2	0.7	0.7	12	8.6	2.3	6-pin direct
Zimo	MN180N18	MN180N18	4	4	0.7	0.7	13.3	9.5	2.6	Next18 direct

GLOSSARY

Address The unique ID assigned to a specific decoder, either for the locomotive or, in an accessory decoder, for the output. For locomotive decoders there are two types of address: 2-digit address, from 1–127, also known as the short address, and 4-digit address, from 128–9999, also known as the long address.

Alternating Current (AC) Alternating current is usually shown as an undulating line purely because the current reverses direction, alternating from positive to negative. The wave line shows current half in the positive side and half in the negative side. Power using this type of current is the main form of power delivered to both businesses and dwellings. It is also the main current used for domestic appliances.

Ampere (A) or Amp A unit of measure of the rate of electron flow or current in an electrical conductor.

AWG (American wire gauge) The standard way to denote wire size in North America. In AWG, the larger the number, the smaller the wire diameter and thickness. The largest standard size is 0000 AWG and the smallest is 40 AWG.

Back EMF (BEMF) Back electro motive force is a voltage that is opposite in polarity, created by the rotation of a coil in a magnetic field. It can be measured and used as feedback by a DCC decoder. Also known as load compensation or scalable speed stabilisation, in other words, locomotive cruise control.

Bearings Parts that assist the rotation of an object, supporting the shaft that turns inside the machinery.

Blast mode Increases the output current to assist in programming a sound decoder. It also assists in programming decoders with a Keep Alive module (energy storage). Blast mode will programme everything attached to the main line being used to programme.

Booster Amplifies the digital packets created by the command station, delivering the resulting digital signal to the track and layout accessories along with additional current to an area of the layout. A booster allows the layout to be split into separate powered sections.

Bus Wiring network. There are three types that are relevant to a model railway layout: the control bus (the wiring from the command station to the track); the cab bus (the wiring network from the handset to various parts of the layout and then back to the command system, for example, the connection from the command system to the fascia panels around the layout); and the accessory bus (the wiring from a power supply to lighting, points, signals, and so on).

Command station/booster The brains of the system, which not only takes information from the controller and converts it into commands, but also sends both power to the track and the commands to the decoders. The command station is used to translate user-entered CV values into computer values (binary) that all decoders understand.

Configuration variable (CV) The 'pigeonhole' where the decoder stores information. There are many CVs, categorising specific characteristics for the running/control of the locomotive. These can be programmed to suit the running requirements.

Consists (double-heading or multiple-unit) If two or more locomotives are assigned to pull a train, they can be programmed to run as one unit, or a consist/double-header. This is sometimes referred to as multiple-unit (MU). There are two main types of consist used: universal or old-style, which is memorised in the DCC system, with the information stored in the handset not the locomotive decoder, and advanced consist, which is stored in the decoder's memory. There are no limits to the number of models in a consist, but there is a limit of 127 consists per system.

Continuity Continuity is the presence of a complete path for the flow of current. A closed switch that is operational has continuity. A continuity test is a quick check to see whether a circuit is open or closed. Only a closed, complete circuit (one that is switched on) has continuity.

Current Current is the flow of electrical charge carriers; in other words, electrons.

Decoder Small device (also known as a chip), which responds to commands that are specifically addressed to it. Can be set up to operate both locomotives and accessories.

Differential signal Most electrical signals are single-ended, comprising a single wire and ground. Differential signals use two wires that are the inverse of each other – when one swings positive, the other swings negative in equal magnitude.

Digital command control (DCC) System that gives the user the ability to control trains digitally, DCC offers the user the capability to run several trains at the same time more realistically even on the same piece of track, with the bonus of more simplified wiring.

Direct Current (DC) Direct current is usually shown as a flat line current either in positive or negative. The electrical charge flows in a single direction, which will flow through a conductor such as wire or copper tape and also through semi-conductors, insulators, and so on. Direct current has many uses, ranging from the charging of batteries to large power supplies for electronic systems and motors, etc. It can also be used for some real-life railways, especially in urban areas. Direct current can be converted from an alternating current supply through the use of a rectifier, which will allow current to flow in one direction only. The reverse (DC to AC) is also possible, by using an inverter.

Dither Noise added to an analogue signal. Its purpose is to improve accuracy when the signal is digitised. In digital command control, it is noise deliberately introduced into the motor's drive signal.

DPDT (double-pole double-throw) switch A switch consisting of six terminals. Each of the poles can complete two different circuits – in other words, each input terminal connects with two output terminals, and all four output terminals are separate. One example is a toggle switch, commonly used to control appliances and electronic devices such as computer monitors, television sets and point motors on a model railway. A DPDT switch is used to control two separate appliances connected to the supply. On a layout, this means the main track and the programming track.

Electrofrog Also known as Live Frog, Electrofrog points are model railway turnouts with no isolated section; in other words, they are live. The Peco Electrofrog point was originally designed in the days of analogue operation with the ability to power sidings, depending on which way the switch rails were aligned. (*See* also Insulfrog.)

Electron A negatively charged particle that can be bound to an atom. It is one of the three primary types of particle within the atom, along with protons and neutrons.

Fleeting A practice used in speed matching that looks at not only matching those locomotives that will run together in a consist, but also the ability to run any locomotive with any other, irrespective of class.

Flux Any effect that appears to pass or travel through a surface or substance. Flux for soldering is a chemical cleaning, flowing or purifying agent.

Function mapping In simple terms, the practice of altering which function keys will activate which function. It applies to both lighting and sound functions.

Functions Controllable features such as lighting, smoke units, sounds built into the decoder. They are accessed via the function buttons on the controller/handset. The number of functions is dependent on the manufacturer of the decoder.

Gauge The distance between the rails. In the real world, 'standard gauge' rails are 4ft 8½in apart, inside-to-inside. In the model world, HO trains run on track gauge that is 0.625in wide and OO run on 4mm track, scaled down from the real thing. 'Gauge' should not be confused with 'scale', which is the proportion of the model to the full-size item.

Handset The means by which instructions are conveyed to the command station, which then tells the locomotive or accessory what to do. Also known as the throttle, controller or cab, for most systems it is a handheld unit. It is possible to have multiple handsets on one command station so that friends and family can control the layout too.

Hertz (Hz) The number of wave cycles (or frequency) passing through a given point in a second.

Impedance (Z) In electrical engineering, impedance is a measure of the opposition to electrical flow, presented by the combined effect of resistance and reactance in a circuit. Represented by the symbol Z in diagrams, it is measured in ohms. For DC systems, impedance and resistance are the same, defined as the voltage across an element divided by the current (R = V/I).

Insulfrog Also known as Dead Frog, an Insulfrog point has a plastic section in both the acute crossing and the frog, making the two frog rails that form the V electrically insulated. (*See* also Electrofrog.)

Java Model Railroad Interface (JMRI) Project building tools for model railroad computer control. JMRI is intended as a jumping-off point for hobbyists who want to control their layouts with a computer without having to create an entire system from scratch.

JST connectors Electrical connectors manufactured to the design standards originally developed by J.S.T. Mfg. Co. (Japan Solderless Terminal).

Kick Start Kick Start is an optional multifunction decoder feature that provides additional torque to overcome motor and drivetrain stiction when a locomotive starts to move from standstill. It is controlled by CV65.

Leading wheel Also known as leading axle, pilot wheel, pony truck or front bogie of a steam locomotive. It is an unpowered wheelset (wheels and axle) found in front of the driving wheels of the locomotive.

LokPilot The standard DCC decoder manufactured by ESU. There is also an 'Fx' version that is a function-only decoder with no motor control.

LokSound ESU's DCC sound decoder. There is also an 'Fx' version that is a function-only sound decoder with no motor control.

Märklin Digital Built around Motorola parts, the Märklin Digital system offered simultaneous control of up to 80 locomotives and 256 accessories, 14 speed steps and an accessory function that could be used for headlights. It first appeared in 1979 at the Nürnberg Toy Fair. In 2004 a new system was unveiled in collaboration with ESU. This two-way protocol allowed for control of up to 65,000 devices, with up to 128 speed steps and 16 functions. In 2013 the extended mfx+ digital system was developed, providing additional decoder features and allowing feedback with mfx+-equipped locomotives.

Momentum effects Mimic the way real locomotives start and stop due to the load they are pulling.

MOROP An organisation made up of model railway associates and enthusiasts from around Europe. The function of MOROP is to define and maintain railway industry standards throughout Europe, and where possible, the rest of the world. The standards are in relation to scale, track width, current supply, couplings, track and wheelset dimensions.

Motor In simple terms, a rotating machine that transforms electrical energy into mechanical energy. Most electric motors operate through the interaction between the motor's magnetic field and electric current in a wire winding to generate force in the form of torque applied on the motor's shaft. Various types are used in model locomotives, including iron-core, coreless, pancake, ringfield, can, open-frame and 5-pole.

Multimeter Measuring instrument that combines the functions of an ammeter, a voltmeter and ohmmeter, as well as other electrical measuring instruments. Its main function is to measure the characteristics of electric signal.

NEM (Normen Europäischer Modellbahnen) Standards Defined and maintained by the Technical Commission of the MOROP, in collaboration with model railway manufacturers, these standards define model railway scales. They guide manufacturers in creating compatible products and assist modellers in constructing model railway layouts that operate reliably. Mostly scale-specific, the standards cover areas such as suggested grades, turnout radii, wheel profiles, coupling designs and DCC. One fundamental principle in the NEM standards is the favouring of operational reliability over exact scale reduction ratios. For example, wheel flanges tend to be proportionally wider in smaller scales. This is a compromise that is acceptable in order to ensure reliable operation. The NEM standards are similar to the standards and recommended practices defined by the National Model Railroad Association (NMRA) in the USA, but the two are not universally interchangeable. In recent years, MOROP and NMRA have been working more closely together to establish common standards for developments such as DCC. European model railway manufacturers generally follow the NEM standards, while North American manufacturers generally follow NMRA standards.

Next18 The Next18 Standard was created by MOROP for extremely tight applications, such as TT- and N-scale locomotives. Although the NMRA has not embraced the Next18 interface, some manufacturers have. In addition to specifying the electrical interface and connectors, the NEM 662 Standard also mandates the maximum physical size of decoders. A fully compliant Next18 decoder is no larger than 15 x 9.5mm.

NMRA (National Model Railroad Association) The member organisation of the scale-model railway community, which promotes the hobby through education, advocacy, standards and social interaction. It has 18,000-plus members worldwide and provides a unique set of tools and services to help modellers become better at what they do and to support their enjoyment of the hobby. The NMRA sets the standards and recommended practices that guide manufacturers worldwide. NEM and NMRA standards are in the main compatible with each other. However, NEM standards can be more applicable to the dimensions used in the model locomotive production and NMRA covers more of the standards associated with the design and manufacture of the digital components of the model railway world.

Ohm (Ω) The measurement of resistance between two points of a conductor – the equivalent of one volt per one ampere.

Oscilloscope An oscilloscope is an electronic test instrument that shows the varying voltages of an item graphically. Oscilloscopes are usually used in the science, engineering, biomedical, automotive and telecommunications fields. This type of equipment enables the reading of voltage waves with the view of analysing them for amplitude, frequency, time interval, distortion, and so on. The main purpose of an oscilloscope is to assist with debugging of electronic signals and analysis or characterisation of electronic signals.

Packet Command message sent by the command station to the decoder. Thousands can be sent by the system in a second.

PluX22 Newer standard interface than the 21-MTC, adopted by the NMRA and MOROP.

Point Movable section of track, also known as a turnout or switch, which allows trains to move from one line to another. The most common type of point consists of a pair of linked tapering rails, known as points (switch rails or point blades), lying between the diverging outer rails (the stock rails). These points can be moved laterally into one of two positions to direct a train coming from the point blades towards the straight path or the diverging path. A train moving from the narrow end towards the point blades (being directed to one of the two paths, depending on the position of the points) is said to be executing a facing-point movement.

Point motor Device for operating a railway turnout, sometimes also known as a switch motor. There are typically three types used on layouts: solenoid, stall (slow-action) and servo.

Polarity An entity contains two distinct and opposite poles that can either attract or repel each other. The term polarity is commonly used in electricity, magnetism, chemistry and electronic signalling to describe the flow of electrons.

Pole The number of circuits within a switch that are controlled by that switch. A single-pole switch controls only one circuit, while double-pole switches control two circuits. In a motor, the term pole, as in 5-pole motor, represents the number of permanent magnetic poles for the rotating motor.

Power districts Electrically isolated sections, which can be beneficial on large layouts that have many trains running on them.

Power supply (PSU) The module that supplies the power to systems and devices. The term PSU is normally associated with a plug-in mobile-phone-style charger or a laptop-style power supply. Anything larger is generally called a transformer.

Programming The action of changing CVs in a decoder to enable different functions. It is usually done on a dedicated programming track or, in some instances, on the main track.

Programming track A dedicated section of track that is fully isolated from the main track, providing a place to carry out programming safely.

Pulse width modulation (PWM) A 'digital' method of motor speed control. It is used to adjust the pulse frequency applied to the motor, usually for minimum speed or motor-generated noise.

RailCom A bi-directional data technology developed and trademarked by Lenz and found in the NMRA Recommended Practices. It can read data transmitted by a decoder, including speed, motor load, CV values and the address, and is completely backwards-compatible with digital command control.

Register Term associated with older decoders that use the Register Mode Programming Standard. This predates the NMRA standards.

Relco track cleaner An electronic device that claims to improve the operation of a track. It uses a high-frequency signal to break down the resistance of any dirt on the track, allowing current to flow freely.

Resistance (Ω) A force such as friction that acts in opposition to the direction of motion of a body or current and tends to prevent or slow down that motion. Measured in ohms, resistance indicates the degree to which a substance impedes the flow of electric current induced by a voltage.

Reverse loop A section of track that loops back upon itself, allowing a locomotive to travel on the same track in the opposite direction from which it came. Locomotives can enter and exit the loop without the need for a turntable or without having to be uncoupled. 'Wyes' or reversing triangles also come under the category of reverse loops.

Reverse loop module A device that makes the wiring of a reverse loop easier without the use of push buttons or toggle switches to operate as the locomotive traverses the reverse section.

RMS (root mean square) meter Measures the square root of the mean square (the arithmetic mean of the squares of a group of values), thus enabling a more accurate measurement of AC current and, in turn, DCC current.

Routes When using points/turnouts with their associated point motors and accessory decoders, some DCC systems allow the programming of a number of pre-programmed instructions simultaneously. This is known as a route.

RPM (revolutions per minute) RPM is used to measure how fast any machine is operating at a given time.

RRamp meter A special meter manufactured by DCC Specialities, which accurately measures DCC volts and amps, and can also measure AC and DC volts/amps. It was designed to read true root mean square voltage and current values, which are proportional to the power being supplied to the layout.

Scale Models are built to scale, defined as the ratio of any linear dimension on the model to the equivalent dimension on the full-size object or prototype. It is expressed either as a ratio – for example, 1:8 scale – or as a fraction – for example, 1/8 scale.

Servo motor A servo motor is a specific motor that will rotate parts of an installation with high efficiency and greater precision than standard motors. There are three types of servo motor: positional, continuous and linear rotation. A servo motor does not rotate like a standard DC motor, and only rotates 180 degrees. Servo motors can be found in robotic arms, legs or rudder control systems, toy cars and model railway signals, crossing gates, and so on.

Soldering The use of a metal alloy to bond other metals together using a soldering iron and solder.

Speed steps Incremental steps on a handset, which go from 0 to full speed. They are also programmed into the decoder. There are three groups of speed steps used: 14, 28 or 128. The most common are 28 and 128. The greater the number of speed steps, the finer the speed control from the handset.

Stall current The current drawn by the motor at locked rotor condition. It is the highest current a motor can draw and is proportional to its rotor resistance. If stall current is drawn by a motor for a long time, the motor will overheat and the windings will be damaged.

Stiction An informal contraction of the term 'static friction', influenced by the verb 'stick', used to describe the situation when two solid objects are pressing against each other and not sliding. An amount of force parallel to the surface of contact is required in order to overcome it.

Swiss mapping Created to be a light mapping process for various countries and railway companies. It is applicable to Zimo decoders only. Zimo calls it the 'Swiss Army Knife' of function mapping.

Terminal block A modular block with an insulated frame that secures two or more wires together. Also known as a connection terminal, a terminal connector or, more informally, a choc block.

Terminator Cable terminations make physical and electrical connections between the cable and the terminal of the equipment, junction or another cable, thereby facilitating the flow of electricity in the desired manner. The terminator is usually placed at the end of a transmission line or daisy-chain bus and is designed to match the AC impedance of the cable and hence minimise signal reflections and power losses.

Throw In switch terms, throw indicates the number of possible output connections that can be made. In point/turnout terms, it refers to the turnout being set to the diverging route.

Tin/tinning The process of using a soldering iron to melt solder around a stranded electrical wire. Wires are often tinned before soldering to keep the fine strands together.

Torque The measure of force required to rotate an object about its axis; this force is usually found in a twisting or turning motion. Torque can also be looked at as the force required to rotate, say, a gear or shaft in order to overcome the turning resistance.

Trix Selectrix Trix is the brand for Märklin two-rail products, which often support NMRA DCC. In Europe, the Selectrix system is supported. When Märklin took over Trix the focus changed to Märklin's two-rail digital for command control. Selectrix is based on a data communication protocol originally developed by Siemens for communications between mainframe computers. It is technically more advanced than NMRA DCC, with a bus system that is much faster and decoders small enough for Z scale, but it is available only from a single source. Döhler & Haas are the sole supplier of the Selectrix integrated circuits needed to manufacture the system's components. DCC manufacturers have the advantage of not being tied to one supplier for a critical component.

Voltage (V) The pressure from an electrical circuit's power source that pushes charged electrons (current) through a conducting loop, enabling them to do work such as illuminating a light.

Voltage drop Voltage drop is the decrease of electric potential along the path of a current flowing in a circuit. Voltage drops in the internal resistance of the source, across conductors, contacts and connectors, are undesirable because some of the energy supplied will be dissipated. The voltage drop across the load is proportional to the power available to be converted in that load to some other useful form of energy.

INDEX

address 8, 17, 21–22, 85, 93, 97, 104–106, 117–121, 123–127, 129–130, 139, 152, 156
Alternating Current (AC) 8, 11, 13, 15, 18–19, 29–30, 32, 42, 49–50, 52, 67, 152–153, 156–157
amp 21, 24, 29, 30, 32, 43, 49–50, 86–87, 108, 140–152
ampere 49, 152, 155
AWG 29, 152

back EMF (BEMF) 21, 86, 88–89, 101, 103, 105–106, 108, 119–120, 123, 132, 152
bearings 62, 66, 73–74, 152
blast mode 21, 119, 152
booster 13, 15, 20, 22, 28–29, 45, 80, 130, 152
bus 20, 30–32, 43, 85, 124, 126, 128, 152, 157–158

command station 11–17, 20–21, 28, 80–81, 84, 100–101, 119–121, 125–126, 152, 154–155
Configuration Variable (CV) 21–22, 85, 89, 93–98, 100–103, 117, 119–123, 125, 130–132, 138–139, 152, 154, 156
consist 10, 21, 97, 100–101, 103–104, 106, 119, 120–121, 129, 152–153, 156
continuity 51–54, 109, 112–113, 128–130, 153
current 8, 13, 15, 19–21, 32, 42–43, 47–49, 51–53, 76–77, 85–89, 105, 108–109, 113, 117, 123, 125, 128, 130, 132, 140, 152–158

DCC 6, 9, 10–28, 30, 32, 39, 40–42, 44–45, 52, 71, 74, 80–109, 114–115, 117–121, 123–127, 129–132, 134, 139, 152–158
decoder 11–12, 15–17, 19–22, 28, 30, 32, 39–40, 42–43, 71, 85–127, 129–132, 134, 138–150, 152–158
differential signal 19, 153
digital command and control 11–16, 153
Direct Current (DC) 8, 13, 15, 17–19, 21–24, 29, 32, 39, 49–52, 108–110, 117, 119–120, 125, 131–132, 153–154, 157
dither 96, 153
double-heading 101–102, 119–120, 152
DPDT 85, 118–119, 153

Electrofrog (Live Frog) 24–25, 153
electron 47–50, 152–153, 156, 158

fleeting 104, 153
flux 54, 56, 61, 153
front bogie 154
function mapping 98–99, 139, 153
functions 21, 86, 90–93, 97–98, 104–105, 108–109, 123, 140, 153–156

gauge 12, 17, 21, 25, 29, 33–34, 36, 39, 43, 46, 86–87, 90, 95, 106, 108, 110, 152, 154

handset 15, 203
hertz (Hz) 19, 50, 154

impedance 154, 157
Insulfrog (Dead Frog) 23–25, 154

Java Model Railroad Interface (JMRI) 22, 135, 154
JST 90–91, 115, 145, 154

kick start 95–96, 106, 120, 154

leading wheel 64, 154
LokPilot 138, 143–144, 154
LokSound 93, 114, 138, 154

Märklin Digital 154
momentum 93, 95, 103, 105–106, 119, 130–131, 154
MOROP 33, 154–155
motor 16–17, 19–21, 27, 75–78, 86–89, 92, 95–98, 104–106, 108–113, 117, 119–121, 123, 125, 130, 132, 154, 156–157
multimeter 19, 51–54, 77, 86, 109, 112, 128–129, 155
multiple-unit 10, 119, 152

NEM 33–34, 106, 155
Next18 91, 104, 140, 143–145, 149, 150–151, 155
NMRA 6–10, 11, 19, 22, 34, 92, 98, 109, 115, 121, 123–126, 130, 134, 155–158

ohm 51, 93, 115–117, 155
oscilloscope 18, 19, 155

packet 155
PluX22 91, 140, 143, 145–146, 149–151, 155
point 15–16, 22, 24–27, 32, 40, 124–126, 153–157

polarity 8, 11, 19, 24–25, 40–41, 49, 120, 125, 152, 156
pole 87, 118, 153, 155–156
pony truck 66, 154
power district 20, 29, 156
power supply (PSU) 12–13, 15, 28–29, 32, 40–41, 45, 50–51, 80–81, 88, 126, 131, 152, 156
programming 21–22, 84–85, 100, 105, 117–119, 121–127, 129–130, 132, 152–153, 156–157
Pulse Width Modulation (PWM) 18–19, 120–121, 156

RailCom 10, 21, 97, 108, 119, 123, 139, 156
register 123, 156
Relco 23, 156
resistance 48, 50–54, 71–73, 77, 129, 154–158
reverse loop 40–41, 156
RMS 19, 52, 156
route 25–27, 30, 132, 157
RPM 88, 157
RRamp meter 19, 157

scale 21, 33–34, 37, 39, 42, 73–74, 89, 95, 106, 108, 111, 116, 119, 154–158
Selectrix 8, 157–158
servo 16, 124–125, 156–157
soldering 30, 47–61, 70–71, 110–111, 117, 153, 157
speed step 93, 95, 97, 100–102, 119–120, 131–132, 139,
stall current 21, 86, 89, 108, 125, 157
stiction 95, 154, 157
Swiss mapping 98, 157
switch 24, 40, 57, 85, 97, 118–119, 125, 128, 153, 155–157

terminal 48, 85, 106, 126, 128, 140, 144, 146–147, 153, 157
terminator 157
throw 118–119, 125, 153, 157
tin 54, 57, 59–60, 157
torque 68, 88, 154, 155, 157
Trix 8, 157
turnout 15–16, 125, 127, 155–157

voltage (V) 8, 11, 13, 15, 19–20, 28–30, 48–49, 51–53, 81, 85, 93, 95, 102, 106, 109, 119–120, 125, 129, 131, 152, 154–158

First published in 2024 by
The Crowood Press Ltd
Ramsbury, Marlborough
Wiltshire SN8 2HR

enquiries@crowood.com

www.crowood.com

British Library Cataloguing-in-Publication Data
A catalogue record for this book is available from the British Library.

ISBN 978 0 7198 4384 6

Typeset by Envisage IT
Cover design by Blue Sunflower Creative
Printed and bound in India by Nutech Print Services

Acknowledgments

I would like to give a huge amount of thanks firstly to my husband, Andy, who has helped me in more ways than one to bring this book to resolution and has been my rock throughout my journey.

Also, many thanks to Mikayla and her grandfather, Peter, who helped with the proof reading and Peter very kindly gave me details on his layout for the scenery section.

Finally, I cannot go without thanking Richard, the engineer at DCC Supplies, who kept me going during the writing of this book. He also allowed me to use photos of the demonstration and testing layout at DCC Supplies for which he has made all the scenery himself.

Sources

The majority of what is written here is derived from my own experience in learning and doing and in running training courses. However, there are many sites online that are very helpful when I want to check something out. I go most often to the following (in no particular order):
DCC Supplies training presentations and handouts;
www.brian-lambert.com;
www.dccwiki.com;
www.nmra.org/standards;
www.railway-technical.com